DINOSAURS
under the BIG SKY

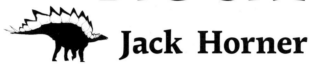

Jack Horner

2001
MOUNTAIN PRESS PUBLISHING COMPANY
Missoula, Montana

All photographs from the author's collection unless otherwise credited

Cover illustration © 2001 by Bill Parsons
Troodon colonial nesting site

Illustrations on pages 64–65, 67, 72–73, and 78 © 2001 by Bill Parsons

Library of Congress Cataloging-in-Publication Data
Horner, John R.
Dinosaurs under the Big Sky / John R. Horner.
        p. cm.
Includes bibliographical references and index.
   ISBN 0-87842-445-8 (pbk. : alk. paper)
1. Dinosaurs—Montana. 2. Paleontology—Montana. I. Title.
QE861.8.M65 H67 2001
567.9'09786—dc21

                                        2001003799

PRINTED IN HONG KONG BY MANTEC PRODUCTION COMPANY

Mountain Press Publishing Company
P.O. Box 2399 • Missoula, Montana 59806
406-728-1900

To my parents,
Miriam and John,
who encouraged me to save stuff

Contents

Acknowledgments   *ix*

1 Introduction   *1*

Amateur Paleontology   *3*

Who Owns the Land and the Fossils?   *4*

Recording Your Discoveries   *5*

The Purpose of Museums   *8*

Monetary Value of Dinosaur Fossils   *10*

Scientific Language and Classification   *11*
  Latin and Greek, The Languages of Paleontology   *11*
  Classification and the Relationships of Dinosaurs   *12*

2 Stories in the Rocks   *16*

Geological Time   *16*

Montana's Geology   *19*

Taphonomy   *28*
  Articulated Skeletons   *30*
  Associated Skeletal Remains   *31*
  Bonebeds   *32*
  Microsites   *34*
  Isolated Bones   *35*

Preservation of Dinosaur Fossils   *36*
  Skeletal Remains   *36*
  Soft Tissue   *37*
  Footprints   *39*
  Eggs and Nests   *40*
  Coprolites and Gizzard Stones   *43*

3 History of Dinosaur Collecting in Montana   *44*

4 Mesozoic History of Montana and Vicinity   *60*

Triassic Time, Scythian Stage   *60*

Jurassic Time, Kimmeridgian Stage   *61*

Cretaceous Time   *63*
    Early Cretaceous Time, Aptian/Albian Stages   63
    Middle Cretaceous Time, Turonian Stage   68
    Late Cretaceous Time, Santonian Stage   69
    Late Cretaceous Time, Campanian Stage   70
    Late Cretaceous Time, Maastrichtian Stage   76
    Latest Cretaceous Time, Maastrichtian Stage   77

5   Dinosaurs and Other Mesozoic Fossils of Montana   81

Jurassic Dinosaurs of Montana   85
    Late Jurassic, Morrison Formation   86
        Ornithischian Dinosaurs   87
        Saurischian Dinosaurs   89
        Nondinosaurian Vertebrate Fossils
           from the Morrison Formation of Montana   92

Cretaceous Dinosaurs of Montana   92
    Early Cretaceous, Cloverly Formation   93
        Ornithischian Dinosaurs   93
        Saurischian Dinosaurs   97
        Common Nondinosaurian Fossils
           from the Cloverly Formation   99

    Late Cretaceous, Two Medicine Formation   100
        Ornithischian Dinosaurs   101
           Hadrosauridae (Duck-bill Dinosaurs)   101
           Pachycephalosauridae (Dome-headed Dinosaurs)   106
           Protoceratopsidae   107
           Ceratopsidae   108
           Ankylosauridae (Armored Dinosaurs)   110
        Saurischian Dinosaurs   111
        Common Nondinosaurian Vertebrate Fossils
           from the Two Medicine Formation   116

    Late Cretaceous, Judith River Formation   117
        Ornithischian Dinosaurs   118
        Saurischian Dinosaurs   122
        Common Nondinosaurian Vertebrate Fossils
           from the Judith River Formation   124

    Late Cretaceous, Livingston Group—Miners Creek,
      Billman Creek, and Hoppers Formations   125

        Late Cretaceous, St. Mary River Formation   *126*
             Nondinosaurian Vertebrate Fossils
               from the St. Mary River Formation   *127*
        Late Cretaceous, Hell Creek Formation   *127*
             Ornithischian Dinosaurs   *128*
               "Hypsilophodontid" Grade   *128*
               Pachycephalosauridae   *129*
               Ceratopsidae   *130*
               Hadrosauridae   *132*
               Ankylosauridae   *134*
             Saurischian Dinosaurs   *135*
             Common Nondinosaurian Vertebrate Fossils
               from the Hell Creek Formation   *138*

6   **Pseudofossils: Dinosaur Bone and Egg Look-Alikes**   *141*
    Concretions   *141*
    Septarian Nodules   *143*
    Geodes   *144*
    Odd Rocks   *144*

7   **Collection, Preservation, and Curation of Vertebrate Fossils**   *145*
    Collection and Preservation   *145*
    Curation   *148*

8   **Museums and Dinosaur Dig Sites in Montana**   *151*
    Montana Museums that Display Dinosaur Fossils   *152*
    Where You Can See Dinosaur
       Fossils Being Excavated   *153*

Appendix I   **Some State and Federal Agencies
                  that Manage Land in Montana**   *154*

Appendix II   **Skeletal Details**   *155*
    General Skeletal Information · *155*
    Skull   *157*
    Vertebrae   *162*
    Front Leg   *166*
    Hind Leg   *168*

Glossary   *170*

Additional Reading   *178*

Popular and Scientific Books that Refer to Montana Dinosaurs   *178*

Books about Dinosaurs from Other States   *179*

Scientific Publications about Montana Dinosaurs   *179*

Index   *189*

Acknowledgments

MY THANKS TO EVERYONE who has ever picked up or discovered a dinosaur or other Mesozoic specimen, and either reported it to a professional paleontologist or donated it to a museum. I couldn't do this without you. I'd also like to thank all the amateur paleontologists who take good, careful notes about their collections and share their data.

I am especially grateful to Pastor Ken Olson of White Sulfur Springs (formerly of Lewistown and Great Falls) for all his specimen donations and collecting efforts over the past couple of decades, and to the following amateurs who, among many others, have made significant discoveries on public lands and reported their finds to the Museum of the Rockies: Larry Boychuk, John Bruninga, Jim and Dennis Coonfare, Norm and Leona Constenius, Buck Damone, Tim Fisk, Dan Holt, Nate Murphy, Sonja and Curt Padilla, Wade Price, Gloria Siebrecht, Doug Tingwell, and a special thanks to Kathy Wankel.

I also recognize and thank amateur collectors Bud Albrecht, David Frasier, and Dan Holt, who have donated specimens found on private land to the Museum of the Rockies and other regional repositories.

Thanks to Canada Fossils Ltd. for its generous donations of specimens collected from the Blackfeet Reservation and for allowing me to see its records of important fossils discovered here in Montana.

A very special thanks to the many landowners who have allowed our teams to collect off their private lands or have given us permission to cross private grounds, including, among many others, the Blackfeet Nation, The Nature Conservancy, Marge Baisch, Big Sky Colony, J. P. Cuny, Denbore Ranch, Scott Frisbee, Jerry Goffena, Junction City Ranch, Wilson Hodgkiss, R. E. Larson, K. Matovich, Huey Monroe, Dan Reading, Ricky Reagan, Gloria Sundquist, Jim and Bea Taylor, Mark and Traci Tilstra, the United Tote Ranch, the Twitchell, Engdahl, Binion, and Trumbo families of Garfield County, and especially Jim and John Peebles and families.

I thank my colleagues who have made significant dinosaur discoveries and reposited them with the Museum of the Rockies, including Mark Goodwin, Des Maxwell, Bill Parsons, Don Rasmussen, Jim Schmitt, Dave Weishampel, and many others.

For all those people who for more than nearly two decades have volunteered their time in the field, among many, many others, Bob Downs, Jay Grimaldi, Barb Haulenbeck, Sid Hoffsteader, Jason Horner, Hilory Korte, Barbara and Robbie Lee, Mark Lindner, Becky Mattison, Pat Murphy, Jill and Phil Peterson, Betty Quinn, Dave Smith, Bea Taylor, and especially Bob Makela.

For logistical help in the field I am grateful to Terry and Mary Kohler, Ricky Reagan, Chuck McAlpine, the Lewis and Vernon Carrol families, Larry and Gail Boychuk, Nels Peterson, and especially Don Dubray.

I am also very grateful to the many people who have worked in the Museum of the Rockies Paleontology Department, including Pat Leiggi, Carrie Ancell, Bob Harmon, Ellen Lamm, Allison Gentry, Karen Chin, Karen Maasta, Jody Smith, Celeste Horner, Pat Druckenmiller, and Jill Peterson. And thanks to the numerous postgraduate, graduate, and undergraduate paleontology students who have discovered and collected a great many dinosaur fossils over the past two decades, particularly Vicki Clouse, Joe Cooley, Kristi Curry, Greg Erickson, Tobin Hieronymus, Jeff LaRock, Des Maxwell, Chris Organ, Ray Rogers, and David Varricchio.

I greatly appreciate the funding of all our field and lab work, especially from Terry and Mary Kohler, Barbara Lee, Nathan Myhrvold, National Geographic, The National Science Foundation, Jim and Bea Taylor, Ted Turner, Catherine Reynolds, James Kimsey, Michael and Donna Coles, Mary Jane Davidson, Universal Studios, and many, many more.

My thanks to the state and federal agencies and other groups that allow us to reposit and care for their collections, including the Army Corps of Engineers, the Blackfeet Nation, Bureau of Land Management, Bureau of Reclamation, United States Department of Fish and Wildlife, and the State of Montana.

I thank Greg Paul for the small drawings of skeletons, Chris Organ for many of the line drawings, Bill Parsons for the color murals, Bruce Selyem for photographic assistance, and Ken Karzminski for historical discussions. Thanks also to the Smithsonian Institution, the American Museum of Natural History, and Bill Clemens for photographs.

Thanks also to Kevin Padian for reading and reviewing the manuscript and suggesting much better ways of saying things. Special thanks to my editor, Kathleen Ort, for her relentless editing, reading, fixing, and straightening out of the manuscript. And thanks to the staff of Mountain Press for publishing this book.

A special thanks to Pat Leiggi, who takes care of all the stuff I hate to do so that I can continue to explore and study the dinosaurs under our Big Sky.

And very, very special thanks to my son, Jason, and my wife, Celeste, for finding a bunch of really cool dinosaurs and for being so patient with this dusty old paleontologist. And thanks to my brother, Jim, my sister, Rosemary, and my mom and dad.

Introduction

WHEN I WAS A YOUNG BOY growing up in Shelby, Montana, I had an appetite for discovery. I'd spend way too much time just walking around with my head down looking for almost anything unusual lying on or sticking out of the ground. Rocks, fossils, money—anything! When I found something sticking out of the dirt or rock, I'd excavate it, and sometimes it would turn out to be a fossil. Shelby sits on black shales of what geologists call the Marias River Formation. The Marias River Formation is part of the Colorado Group, a thick stack of sediments that were deposited in an ocean that existed at the time of the dinosaurs. The fossils I was most likely to discover around Shelby were the remains of sea creatures such as clams, ammonites, and snails. These fossils, together with a bunch of unusual or pretty rocks and other cool junk, were the treasures of my childhood.

A lot of my rock treasures came from deep within the ground. You see, my father was a co-owner of a gravel plant. He and his partner extracted gravel from the places around Shelby where Pleistocene glaciers and rivers had pushed and carried rock down from Canada. Sometimes he excavated treasures for me. My father wasn't a degreed geologist or a geological engineer, but he had a good understanding of geology, and he knew where to look for certain rocks and minerals. He knew, for example, where to look for Pleistocene gravel, and he also knew a few things about the other rocks in north-central Montana. When I was eight years old, he took me to a place near the town of Dupuyer that had once been his ranch. My father had remembered riding his horse across a pasture and seeing large bones sticking out of the rocks. Recognizing my interest in fossil animals, he took me there to explore for dinosaurs. There, in a hillside, I found my first dinosaur bone. It was the beginning of my career, first as an amateur collector and later as a professional paleontologist. I still have that first dinosaur bone—it sits on a shelf in my office at the Museum of the Rockies, where I am curator of paleontology.

As my interest in paleontology grew, my mother, who liked to drive around the countryside, took me to places of my choice in northern Montana where I could explore for dinosaur bones to add to my ever-growing fossil collection.

When I was a senior in high school, I put together a science project comparing the dinosaur remains from the Judith River Formation of Montana with the dinosaur remains of southern Alberta. In hindsight, I realize that many of my identifications of the dinosaurs were wrong. I simply didn't have the books to identify all the bones and pieces of bones, nor to identify dinosaur species based on fragmentary remains. This book is the book I wish I'd had when I was an amateur paleontologist exploring and collecting dinosaur fossils. This book describes the different species of dinosaurs known from Montana, explains the scientific importance of dinosaur bones and skeletons, and discusses how amateur paleontologists can help professional paleontologists. My hope is that the information here will also help you to identify some of the common dinosaur fossils you might happen across while walking around Montana. I wrote this book for both the veteran collector and anyone interested in knowing more about the dinosaurs that once lived under the Big Sky.

Montana is world famous for its dinosaurs. The first dinosaur remains found in the Western Hemisphere came from near the mouth of the Judith River in an area of the Nebraska Territory that is now central Montana. The first dinosaur eggshell reported from the Western Hemisphere came from Montana. So did the first nest of baby dinosaurs found in the world and the world's first dinosaur embryos. The first remains of *Tyrannosaurus, Maiasaura, Deinonychus, Troodon, Ankylosaurus, Orodromeus, Tenontosaurus, Zephyrosaurus, Avaceratops, Einiosaurus,* and many other dinosaurs were also found in Montana. What's more, many of the original discoveries were made by amateur paleontologists, rock hounds, and other people who just happened to be out walking around and looking down at the ground.

Montana has influenced dinosaur research as well. Many significant discoveries here have initiated ideas about the relationship between dinosaurs and birds, about dinosaur social behavior, and even about dinosaur physiology. Montana has played a consequential role in the controversy concerning dinosaur extinction and has shaped ideas about what dinosaurs ate, where they nested, and how they evolved.

Amateur Paleontology

To uncover a dinosaur skeleton, or even a single bone, from the earth and be the first person, or even among the first people, to lay eyes on it, is probably one of the most exciting things you'll ever experience. My excitement after finding that first dinosaur bone on my father's ranch is one of the main reasons I wanted to be a paleontologist. And even now, almost fifty years and thousands of dinosaur remains later, I still thrill at seeing a skeleton in the ground during excavation—even if it's the skeleton of a dinosaur I've seen many times before. The excitement never seems to wear off.

Collecting dinosaur fossils is not only fun but also a great excuse to get outside, enjoy spectacular landscapes, and breathe some fresh air. In Montana, most dinosaur fossils come from areas of badlands, where erosion eats away at hillsides faster than plants can grow on them. Many people consider these badlands desolate and bleak, but paleontologists find them extraordinary and peaceful.

Amateur collectors are very important to museum research programs. In Montana, for instance, a number of amateur collectors spend their free time during the summer directly aiding the Museum of the Rockies with its research efforts or collecting fossils on their own and bringing in important specimens. These dedicated amateurs found many of the display-quality specimens on view at the Museum of the Rockies. Most museums benefit from the contributions of similar amateur collectors. Amateurs often participate in the actual research and are acknowledged when their specimens are described. Amateurs are not paid for specimens, but the museums generally cover their expenses for travel and supplies. Interested people can write or call their local or state museum to find out how to volunteer time to such projects.

Dinosaur paleontology is a team effort, and everyone who enjoys collecting can help acquire information about these incredible, extinct animals. And while finding the remains of dinosaur bones or skeletons is an exciting and healthful activity, both the exploration and the collection must be done responsibly. Irresponsible collecting can do more damage than if the fossil had simply weathered away in the rain. Amateur paleontology, just like professional paleontology, carries some important responsibilities. These include getting permission to collect, keeping good records of where and in what kind of rock you found your specimens, and making the specimens available for scientific study.

Who Owns the Land and the Fossils?

All the land in the United States, including all of Montana, is owned by someone or some organization, and that means that *before you can explore or collect, you must get permission.* It is best to get written permission.

Most dinosaur remains come from badlands, and most badlands don't have enough grass to support grazing animals such as cows and sheep, so most badlands are owned by the federal or state government as public land. The public may use this land, but some restrictions apply. For instance, the archaeological artifacts and vertebrate fossils, including dinosaur remains, found on federal land belong to the people of the United States, and those found on Montana state land belong to the people of Montana. Individuals are not allowed to own artifacts and vertebrate fossils from public land because they belong to everyone.

As collectors, we have a responsibility to care for these items. Collecting dinosaur bones and other vertebrate remains on public land requires a permit, and collected items are to be stored in national repositories. Of course, collecting from national parks is prohibited. The state owns some of the land in Montana. Collectors must obtain a permit from the Montana Department of Conservation and Natural Resources to collect on this land. Collecting from state parks is also prohibited. Appendix I lists state and federal agencies responsible for managing public lands in this region and their contact information.

On private land—land that a private individual or group of private individuals owns—the landowner generally owns the fossils.

It is up to the landowner to give permission to collect and keep fossils from private property. It is important that you not only ask for permission to collect but also show the landowner what you find and express thanks when you are finished. Such courtesy will make return visits much easier. Keep in mind that even if a landowner grants permission to collect and keep fossils on his or her property, the landowner may also revoke that permission.

Most government and state agencies prohibit collecting fossils from public lands for the purpose of building private collections. Personal collections can be legally made from private lands where landowners have given permission. The permits that government agencies grant are generally for people directly involved in research or salvage operations directed by curators of repository museums.

Laws about collecting vary from place to place. For example, the Canadian province of Alberta strictly forbids the collection of vertebrate fossils on any land because all fossil vertebrate specimens belong to the provincial government. Only the Royal Tyrrell Museum of Palaeontology in Drumheller, Alberta, may collect and reposit the province's vertebrate fossils.

Wherever you wish to collect, be sure you have permission from the landowner or managing agency before you begin any kind of exploration or excavation.

Recording Your Discoveries

A paleontologist carries a number of items into the field to collect specimens and scientific information. The most important item is not a digging device, a rock hammer, or any other tool, but rather a notebook. I make a new notebook for each year, and in it I put landowners' names and phone numbers, and all kinds of information about the things I see or find. I make drawings similar to those in the margins of chapter five in this book. You can never have too much information, but you can certainly have too little.

A professional or amateur paleontologist's most important job is to record observations about the geographic and stratigraphic location of the fossil, and then describe the rocks and the fossil itself while it is in the ground. Your first job is to determine exactly where the fossil is. The best way to do this is to pinpoint the location with a Global Positioning System, or GPS. If you don't

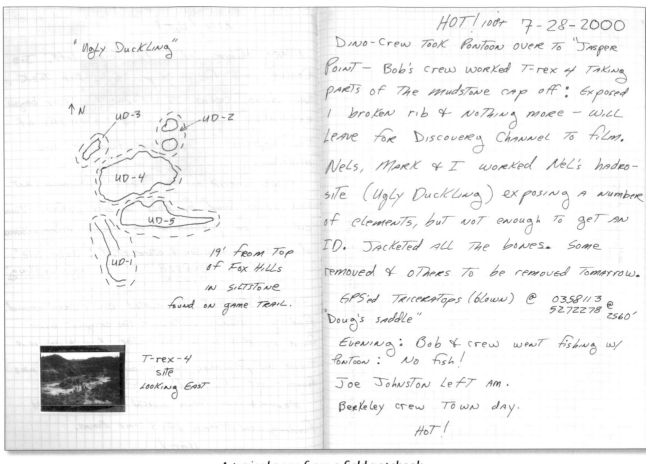

A typical page from a field notebook

have a GPS, you can find your location on a topographic map, or at least determine how far and in what direction the specimen is from some well-known landmark.

When you've determined where you are geographically, you must figure out where you are stratigraphically. Stratigraphy is the term we use when talking about layers of rock. Through time, sediments pile on top of one another, much like layers in a cake. Together these layers make up the stratigraphy of the area, so one of the pieces of important information about the fossil you have found is determining where the fossil is in relationship to the rest of the stratigraphic layers. We determine stratigraphic position by measuring up or down from the fossil. Geologists refer to this kind of measuring as "measuring the section"—the section being all the vertical strata that can be measured in a particular area. If the fossil is on the side of a hill, a person measures the vertical distance from the bottom of the hill to the top of the hill, describing where in that vertical section the fossil sits.

Other people can find their way to your site by using their GPS unit to pinpoint the geographic location you gave for the hill. Knowing where the fossil is vertically will allow them to find the specimen, or the place where you excavated it. Even if there is no evidence of exactly where the specimen was, your stratigraphic description will let them know its approximate layer. Stratigraphy helps paleontologists and geologists determine relative ages of specimens. If, for example, you found a number of specimens 20 feet above a certain layer of coal or other distinct layer of rock, then we can hypothesize that the fossils represented animals living at about the same time the coal was deposited.

One of the easiest methods for measuring a section is by sight. Simply measure how far it is from the ground to your eyes. When you walk up to a hillside, you can pick a spot horizontal to your eye level. Then climb the hillside, placing your feet at the spot you picked. Then find another spot at eye level, and so on until you reach the top of the hill. Add up the number of spots you picked, multiply by your eye height, and you will have the approximate height of the vertical section. The most important part of this exercise is to be sure the rock layers are lying flat. If the sediments are tilted, you have to measure the section along a line that is perpendicular to the layers—we are much more interested in where the fossil is with relationship to the sedimentary layers than to the height of the hill.

A typical geographic and stratigraphic location record would read like this:

45° 35.118' N, 118° 55.145' E
(latitude and longitude coordinates)

NE facing hillside 20 feet up from base of hill in brown sandstone.
(Local stratigraphic section is 80 feet thick. Fossil site is 60 feet from bottom, 20 feet from top.)

When you have determined where you are geographically and stratigraphically, record this information in your field notes and give the location a field number. I generally assign a number that refers to me as the discoverer (JRH), the date that I found the specimen (7-14-01), and then a number to separate it from other discoveries that day (6), so my field number would be JRH 7-14-01-6. If I collect the fossil, I wrap it in toilet paper, in paper towel, or in a plaster jacket, layers of plaster-soaked fabric that protect the fossil in much the way a plaster cast protects a broken arm. I write the field number on the package using a permanent marker and also typically put a label with the field number in the bag or jacket in case the outside number gets rubbed off or the bag tears. The field number is the only record you have of the specimen, so it is very important that this data remain with the fossil.

After determining where the fossil is geographically and stratigraphically and assigning it a field number, make a drawing of your discovery. Draw it first as you find it and second as you excavate it. If it's a single bone lying on the surface, the excavation drawing is not necessary. But if the specimen appears to be several associated bones, and you have to excavate them, it is important to draw a picture of how the bones are lying in relationship to one another. This information, along with descriptions of the kind of rock that holds the specimen, might allow future scientists to determine something about how the fossils got to their location and maybe even something about the environment in which the animal died. Make the drawing in your notes, and note the color of the bones and any other interesting features of the sediments. Maybe you notice burrows that insects made, plant debris, or mollusk shells. If there are lots of interesting features or a number of bones together, take a photograph. All this information is useful to scientists and will be greatly appreciated when you share this information.

The Purpose of Museums

Museums are not in the business of owning fossils or artifacts, but rather endeavor to save them for future generations of the enthusiastic public and for both amateur and professional scientists. Research-based museums, like the Museum of the Rockies, are in the business of collecting fossils primarily to learn new information about extinct creatures and to display this information in such a way that everyone can learn about our region's past. The Museum of the Rockies is part of Montana State University, and as such is technically one of Montana's state museums. The majority of the money used to operate the Museum of the Rockies comes from donations rather than from the state. That means the museum depends on private donations of both specimens and funding. Most of the joint repository/display museums in the United States are operated in association with universities and rely on donations.

The best thing you can do with an important specimen to ensure its safety and scientific value forever is to donate it to a museum. Specimens that collectors pick up and haul home might be important to science, but they are of little or no scientific value if nobody knows about them. Over the years, I've met people who claim to have found extraordinary fossils, but they are not interested in sharing them with the public or with science. If the fossils are extraordinary, perhaps rare or even new to science, it's a shame to hide them away. After a few years you might forget about them or even lose them. And almost always the important information about their location in the field is eventually lost. I encourage anyone who thinks they might have a potentially interesting fossil to take it to a paleontologist or a museum and have somebody take a look. If the paleontologist or other museum representative determines that your specimen might be important, seriously consider donating the specimen so it will be both safe and available for study. Museums are places where we keep things so they don't get lost.

If you desire to keep a private collection at your home, I recommend taking a few steps to prevent the loss or destruction of the collection. Do not rely on your memory! Keep records of all important specimens. Number and reference each item in some kind of catalog. Then, if something happens to the original collector, others will be able to determine the locations and identifications of all the specimens. In addition, register a copy of your files with a local or state museum. That will allow interested professionals to find out about possibly important specimens and be better prepared to care for the fossils if you lose interest in them. When you register or donate your collection, the collecting records will always show that you collected the specimens. Your name will be attached to those specimens forever, and when paleontologists write about them, they will acknowledge your contribution.

Monetary Value of Dinosaur Fossils

Some individuals and companies collect dinosaur fossils as a way to make money. As we saw in the late 1990s with the sale of the *Tyrannosaurus rex* named Sue, some dinosaur specimens sell for a great deal of money. Such high selling prices are unusual, though, because most museums, particularly those in the United States, don't have the money to purchase dinosaur skeletons, and wouldn't even if they could. It is not in the interest of science to fuel the fossil market—buying fossils encourages people who know little or nothing about the science to go out and collect fossils for profit. Furthermore, most museums have their own paleontologists to gather fossils for study and display.

The Museum of the Rockies does not buy or sell vertebrate fossils and does not encourage anyone else to do so either. Instead, we try to encourage people to donate their fossils to a museum where the public and scientists can enjoy and study them. I donated the collection I made in high school to Princeton University. It is now under the care of the Peabody Museum at Yale University in Connecticut. When I was looking for a place to keep my collection safe, there was no repository museum in Montana with adequate space. That the collection is no longer in Montana is of little consequence because I can always borrow the specimens. The point is that I know where the specimens are, and I know they are safe.

In contrast, most of the dinosaur specimens that commercial collectors collect in Montana and later sell are now either in private collections or in museums around the world. I think it's important to keep scientific treasures collected from the United States in the United States and accessible to science and the public. Most other countries treat their fossils in the same way.

Because large sums of money are often involved in the commercial collection of fossils, it becomes difficult to determine which commercial collectors are scrupulous. When the prime reason for collecting a specimen is to make money, the accuracy of the data is always in question. For purposes of scientific research, the data about every fossil must be reliable. Falsely documented or composite specimens cause major problems with interpretations. One of the biggest problems with commercially collected specimens is that there is often no way to know for sure where the fossil was found, what the geological context was, or whether the specimen is made from the bones of one animal or several— or even from the bones of more than one species. An example in 2000 involved a "chimaera," or composite skeleton, from China. The creative commercial collectors apparently combined a bird skeleton with a different dinosaur skeleton and manufactured a new species of feathered dinosaur. This new species appeared so important that *National Geographic* published a story about it. When later investigation revealed that the specimen was fake, it cast doubt on the authenticity of all the specimens discovered from that area. Scientists no longer trust the commercially collected specimens coming out of that area of China.

Many people now making collections to sell have little or no formal training in paleontology. Many of these folks don't know what kinds of information are important. As a result, they often destroy or lose the information. The most reliable data come from amateur and professional paleontologists who collect for the purpose of acquiring new information rather than making money.

Scientific Language and Classification

In all fields of science, researchers use specialized words to convey technical information and specialized methods to get certain kinds of data. Dinosaur paleontology is no different. It has its own unusual terms and methods that draw from both biology and geology. To be able to collect and document fossils, you don't need to know all the terms and methods paleontologists use, but you should understand some of them. The words you'll want to know include the names of the bones of a skeleton and the terms used to describe particular rocks and groups of rocks. You should also know how to collect a fossil without causing its destruction.

Latin and Greek, The Languages of Paleontology

All paleontologists around the world write the scientific names of all plants and animals, and even some names of bones, in Latin or Greek rather than English. The classical languages have been the universal language of all sciences since the Renaissance. This is because in Europe, where many of these sciences developed, the common languages of scholars were Greek and Latin, regardless of what language they spoke at home. This ensured that everyone used the same names for the same organisms.

The scientific name of an organism has two parts: the genus and the species. The scientific name for humans is *Homo sapiens*. In Latin, *Homo*, the genus name, means "same"—that is, the same animal as us, humans—and *sapiens,* the species name, means "wise." Linnaeus interpreted *Homo sapiens* as "know thyself," quoting one of the great Greek philosophers. The scientific name for *T. rex,* of course, is *Tyrannosaurus rex. Tyrannosaurus* is Greek for "tyrant reptile," and *rex* is Latin for "king." Another rule in scientific names is that we write both the genus and species names in italics. In this and all other scientific books, all genus and species names are treated this way. When we are writing by hand— for instance, in our field notebooks—we underline words that should be in italics.

Paleontologists follow classical anatomists in using Latin for the names of the bones in the skeleton. In Latin, we indicate singular (one) and plural (more than one) in a different way than in English. For instance, one thighbone is called the femur; if we have more than one, we call them femora. The shinbone is called the tibia, and its plural form is tibiae. A single bone from the backbone is called a vertebra, and a bunch of them are vertebrae. A word in Latin that ends in "ae" often means more than one.

The glossary near the end of this book lists unusual terms and their meanings. I define terms the first time they appear in the main text. If you need to review the meaning of particular words, check the glossary.

In paleontology, we measure skeletons and bones using the metric system. The line below is 10 centimeters (about 4 inches) long.

---

10 centimeters

If you are interested in the technical descriptions of the bones of dinosaur skeletons, you can refer to Appendix II, Skeletal Details. There you can learn about all the bones of the skeleton and how they connect to one another.

Classification and the Relationships of Dinosaurs

Over the past twenty years or so, scientists have been learning that the system we have used to classify living and fossil organisms, including dinosaurs, is out of date. The old classification system, known as the Linnaean Classification System, was based on grouping organisms in ranks according to similar characteristics. Carl Linnaeus created this system in the late 1700s before scientists knew that all organisms had evolved from other organisms and that most organisms that have ever lived are now extinct. The ranks include the main categories of kingdom, phylum, class, order, family, genus, and species. According to this older system, an ostrich is in the Kingdom Animalia, Phylum Vertebrata, Class Aves, Order Ratites, Family Struthionidae, and Genus and Species *Struthio camelus*. The Linnaean system has no way to show how one organism is related to another as a result of evolution—that is, descent with modification. For this reason, the Linnaean system is now considered obsolete.

A new classification system, called cladistics, attempts to do away with the ranks of kingdom, phylum, and so on. Instead, it organizes plants and animals by how closely they are related to one another, based on characteristics acquired through evolution. To make a long story short, one of the things we've learned is that some groups in the Linnaean system—for instance, the Class Reptilia—included some animals that weren't related to others in the same group. According to the Linnaean system, all the vertebrate animals that lay eggs on land, except mammals and birds, belong in the group Reptilia. But we now know that the animals we used to call "mammal-like reptiles" do not share direct ancestry with such animals as turtles, lizards, and snakes, and therefore didn't belong in the same group. In the new cladistic system, we have placed all the animals more closely related to mammals than to turtles into a group called Synapsida. Turtles, together with a group called diapsids that includes lizards, snakes, crocodiles, and birds, make up the new group Reptilia. So, the new grouping of reptiles includes turtles, lizards, snakes, crocodiles, pterosaurs, dinosaurs, and birds. Birds? Yes, birds. Birds are a group within the Dinosauria. All dinosaurs are related to one another and therefore share a common ancestor. Birds are related to dinosaurs by a common ancestor thought to be a small coelurosaurian dinosaur related to *Velociraptor* and *Deinonychus*. Since dinosaurs and birds share a common ancestor, and since dinosaurs are reptiles, then so are birds. That means ostriches are actually a group of reptiles. Because birds are dinosaurs, we have to distinguish the living birds from the extinct dinosaurs. We now call the extinct dinosaurs the nonavian dinosaurs, and the living birds the avian dinosaurs. That means this book should really be called *Nonavian Dinosaurs under the Big Sky.*

Cladograms are diagrams that show relatedness. Take a look at the accompanying cladogram showing the evolutionary relationship of mammals and reptiles. The group Reptilia, which includes turtles, lizards, snakes, crocodiles, and dinosaurs, shares a common ancestor with mammals. Mammals didn't evolve from reptiles, but instead they simply shared an ancestor. Crocodiles share an ancestor with Dinosauria. When two branches of a cladogram come together, it indicates that the two groups are related by ancestry.

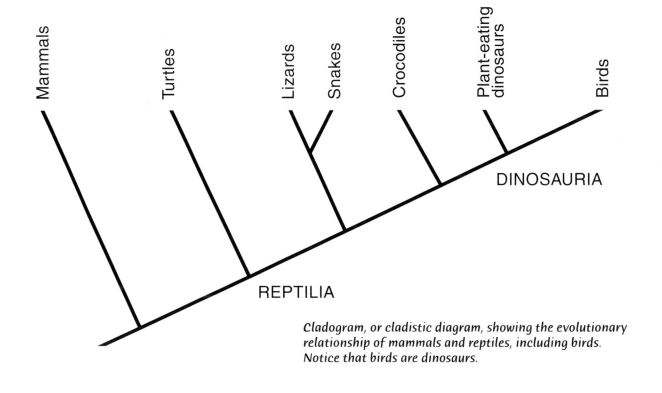

Cladogram, or cladistic diagram, showing the evolutionary relationship of mammals and reptiles, including birds. Notice that birds are dinosaurs.

There are two main groups of dinosaurs: the Saurischia and the Ornithischia. The Saurischia include the bipedal, meat-eating theropods, such as *Deinonychus, Tyrannosaurus,* and all birds, and the long-necked, plant-eating, small-headed sauropods, such as *Diplodocus* and *Apatosaurus.* The Ornithischia contains a variety of plant-eating dinosaurs including primitive ornithopods, such as *Zephyrosaurus* and *Orodromeus,* and more advanced ornithopods, such as the duck-bills *Gryposaurus* and *Edmontosaurus.* Other ornithischians include the pachycephalosaurs, or bone-heads, including *Stegoceras,* their cousins the ceratopsians, or horned dinosaurs, such as *Leptoceratops* and *Triceratops,* and the thyreophorans, or armored dinosaurs, such as *Edmontonia* and *Ankylosaurus.*

In Mesozoic sediments under the Big Sky, we have examples of most of the different groups of dinosaurs, but only from rocks of late Jurassic and Cretaceous age. Montana's Triassic rocks have not yet produced any dinosaur remains.

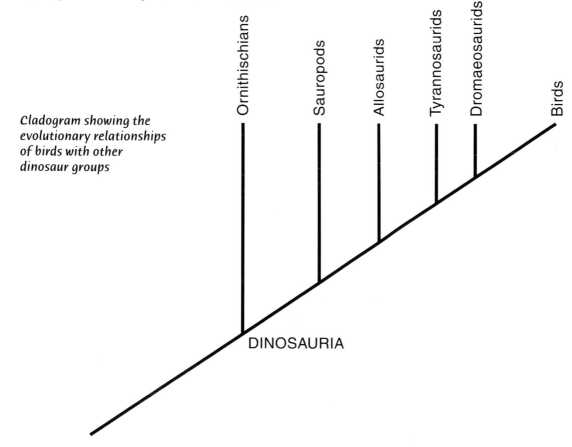

Cladogram showing the evolutionary relationships of birds with other dinosaur groups

Stories in the Rocks

PALEONTOLOGY IS A COMBINATION OF BIOLOGY AND GEOLOGY—it is the study of extinct organisms that are preserved in rocks. Knowledge of various fields of biology, including anatomy and histology, are important in studying dinosaurs, but only after the fossils have been collected and are freed of all the sediment that encased them. Geology is the most important science for a dinosaur paleontologist to know because we find dinosaur skeletons in rocks. Our knowledge of geology helps us understand where to look, what to look for, and how old the fossils are. Geological information is essential because it helps paleontologists figure out what happened to animals, what may have killed the animals, and what happened to their remains after they died. Geology tells us the stories in the rocks.

Geological Time

The geological rock record of our earth represents about 4.6 billion years of geological and paleontological events. To keep track of where they find particular fossils within the 4.6-billion-year rock record, geologists and paleontologists have devised a geological time scale. The time scale breaks the earth's long history into units similar to the hours, minutes, and seconds on a clock. On the geological time scale, we call the units of time (from longest to shortest) eons, eras, periods, epochs, and stages. Dinosaurs lived through most of the Mesozoic Era of the Phanerozoic Eon. Within the Mesozoic Era are three periods: the Triassic, Jurassic, and Cretaceous Periods. Each of these periods is broken into several epochs and stages. This book uses the names of some stages, but not the epochs. *Tyrannosaurus rex* lived during the Maastrichtian Stage of the Cretaceous Period of the Mesozoic Era of the Phanerozoic Eon. But for our purposes, we can just say that *T. rex* lived

GEOLOGICAL TIME SCALE				millions of years ago
Eon	Era	Period	Stage	
PHANEROZOIC EON	Cenozoic	Quaternary		
		Tertiary		
	Mesozoic	Cretaceous	Maastrichtian Campanian Aptian/Albian	64.5
		Jurassic	Kimmeridgian	144
		Triassic	Scythian	208
				230
	Paleozoic	Permian		
		Pennsylvanian		
		Mississippian		
		Devonian		
		Silurian		
		Ordovician		
		Cambrian		
				570

The geological time scale, showing the eras and periods of the Phanerozoic Eon and the stages of the Jurassic and Cretaceous Periods from which dinosaur remains are known

during the Maastrichtian Stage. The most important stages to remember for Montana are the Maastrichtian, Campanian, and Aptian/Albian Stages of the Cretaceous Period, and the Kimmeridgian Stage of the Jurassic Period. These are the geological stages during which most of the dinosaurs of this region lived.

How do we really know how old a dinosaur fossil is? Dinosaur fossils cannot be dated directly. They are too old for the carbon dating techniques that scientists use to date more recent fossils and archaeological artifacts, because most carbon 14 decays to carbon 12 after about 70,000 years, and our dinosaurs are about ten thousand times older than that. To bracket the age of dinosaur fossils, the paleontologist has to date the rock in which he or she finds the fossil or one that is relatively nearby. A researcher can estimate a general date from knowing the age range of the geological formation that contains the fossil. For example, potassium-argon dates tell us that the Judith River Formation is between 80 million and 74 million years old. (Potassium isotopes decay to argon far more slowly than carbon isotopes decay, so paleontologists find these elements useful in dating dinosaurs.) If you find a skeleton near the bottom of the formation, you could guess that the specimen is probably around 80 million years old. If one came from the middle of the formation, you could hypothesize that it was around 77 million years old. But if you needed a more precise date, you'd have to find a bed of volcanic ash, which contains elements that can be dated directly, within the formation and near the level of your fossil. Then you would send a sample of the ash to a dating laboratory.

I send most of the rocks I want dated to a laboratory in Berkeley, California. At that laboratory, scientists use precise machines to measure the amounts of particular radioactive elements in a rock to determine the rock's age. Over time, a radioactive parent element decays into a lighter daughter element. Half the atoms of any amount of a radioactive parent isotope will break down into its daughter element in a set period of time called the radioactive element's half-life. By measuring the proportions of radioactive elements and their daughter products, researchers can determine how many years have passed since the rock formed. The half-life, then, can serve as a clock to help us date rocks. For example, fresh volcanic ash contains the radioactive parent isotope potassium 40. Potassium 40 decays to form the daughter isotope argon 40. The half-life of potassium 40 is about 1.3 billion years—half of the original potassium changes to argon 40 in 1.3 billion years. Measuring the amount of argon 40 in volcanic ash associated with a dinosaur skeleton gives a relatively accurate date for when the volcanic ash spewed out of the volcano—and when sediment buried the dinosaur skeleton. Potassium-argon dating was very popular a couple of decades ago, but with ever more precise machines, most age analysts now use the isotopes argon 40 and argon 39. These isotopes give a more precise date to the rocks, which means that the potential error is smaller.

The Egg Mountain dinosaur site near Choteau has been age dated as 76.7 million years, with an error factor of plus or minus 250,000 years. From the perspective of a dinosaur paleontologist, 250,000 years is almost no time at all.

Montana's Geology

Dinosaur remains are common in Montana and Alberta because most of the rock from the time of the dinosaurs now sits at the surface, where wind and rain can erode it. Elsewhere, including in much of Wyoming, most of the rock of the right age is still buried under thousands of feet of younger rock. In still other places, such as Minnesota, rocks of the right age have long since eroded away. There, the rocks exposed at the surface are either too old for dinosaurs or too young. We are fortunate that most of the eastern parts of Montana and Alberta consist of rocks from the Mesozoic Era, the age of the dinosaurs.

Why do dinosaur skeletons seem relatively common in the fossil record while skeletons of modern animals, such as cows, are so rare out on the prairie? The reason is simple. Skeletons can only be preserved if they get covered by sediments before they rot away. Prairies are areas of erosion rather than deposition. On the prairie, there is no way for sediments to bury a carcass. So it gets eaten, strewn around, or simply rots away. The skeletons we find in the fossil record were covered before they were eaten or rotted. The geological layers in which we find dinosaur skeletons all are sediments that wind, streams, rivers, lakes, or oceans laid down. These are areas of deposition rather than erosion. There are no prairie environments preserved in the fossil record.

We find fossil plants and animals in sediments that were once soft and muddy but have since hardened into rock. In the northern Rockies, we find most skeletons in rocks that were deposited originally as river sand and mud. When a dinosaur died either in a river channel or along the river bank, sediment buried its body. The dead animal's soft tissues rotted away, leaving just its skeleton. Over thousands or even millions of years, more sand and mud stacked up over the skeleton, in some places piling as much as several miles thick. After more time, other geological processes, such as earthquakes and mountain building, pushed the rock layers containing the dinosaur's remains close to the surface. There erosion stripped away the overlying rock. After millions of years, the skeleton began to erode out of its rock tomb.

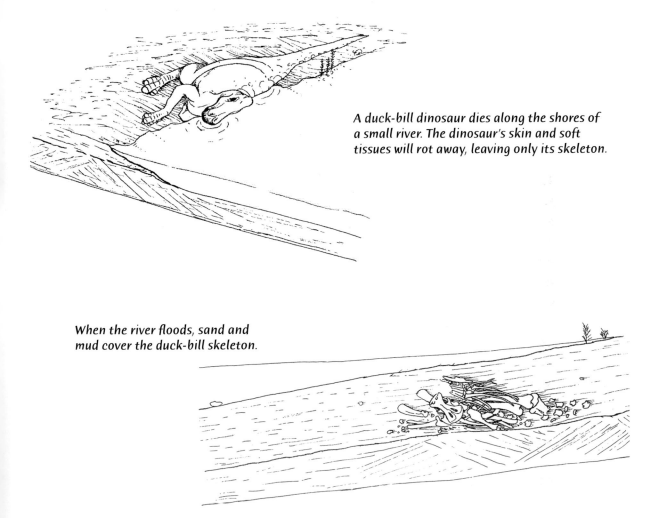

A duck-bill dinosaur dies along the shores of a small river. The dinosaur's skin and soft tissues will rot away, leaving only its skeleton.

When the river floods, sand and mud cover the duck-bill skeleton.

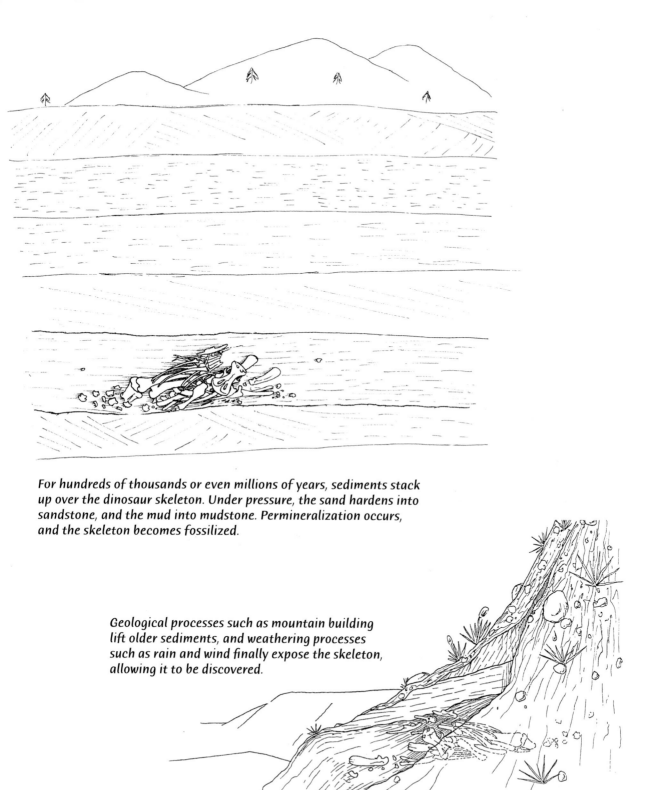

For hundreds of thousands or even millions of years, sediments stack
up over the dinosaur skeleton. Under pressure, the sand hardens into
sandstone, and the mud into mudstone. Permineralization occurs,
and the skeleton becomes fossilized.

Geological processes such as mountain building
lift older sediments, and weathering processes
such as rain and wind finally expose the skeleton,
allowing it to be discovered.

When geologists talk about rock layers that contain fossils, they typically use the word *formation*. A geological formation is a rather large package of sediments that look similar and contain a particular group of fossils. A formation is also defined as covering a geographic space large enough to be mapped. This means that it usually covers at least several hundred square miles.

In Montana, we have numerous sedimentary formations that are of Mesozoic age. Some of them consist mainly of marine sediments and contain few, if any, dinosaur remains. Others contain mainly sediments deposited in nonmarine environments such as rivers and lakes. Most important of the terrestrial formations in Montana are the late Jurassic Morrison Formation, the early Cretaceous Cloverly Formation, and numerous formations of late Cretaceous age. The fossils we can collect in a particular area depend on which sediments are exposed at the surface of the ground.

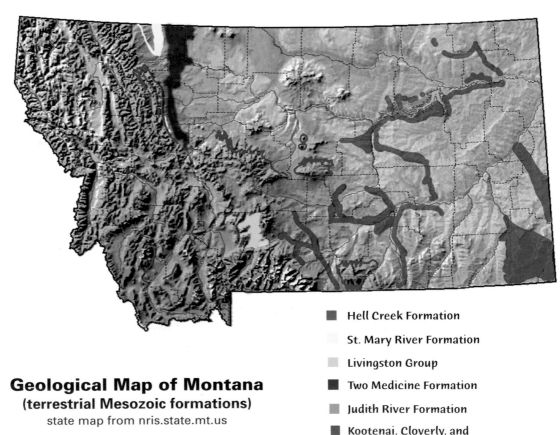

Geological Map of Montana
(terrestrial Mesozoic formations)
state map from nris.state.mt.us

■ Hell Creek Formation

   St. Mary River Formation

   Livingston Group

■ Two Medicine Formation

■ Judith River Formation

■ Kootenai, Cloverly, and Morrison Formations

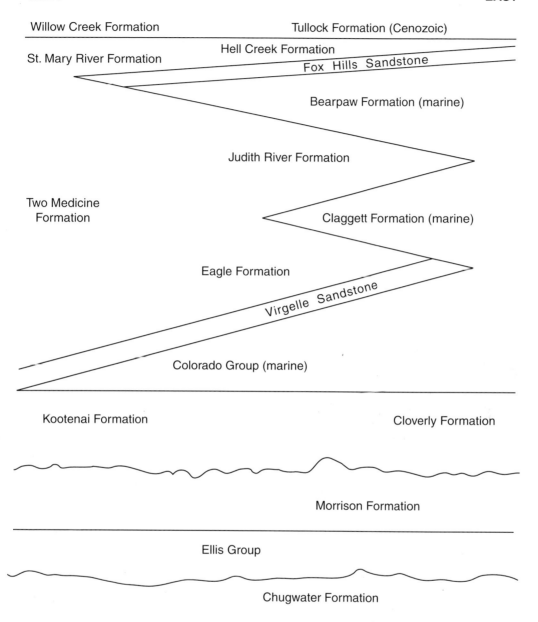

WEST   EAST

Willow Creek Formation   Tullock Formation (Cenozoic)

                                          Hell Creek Formation
St. Mary River Formation   Fox Hills Sandstone

                                          Bearpaw Formation (marine)

                        Judith River Formation

Two Medicine
Formation   Claggett Formation (marine)

            Eagle Formation

                        Virgelle Sandstone

                        Colorado Group (marine)

    Kootenai Formation   Cloverly Formation

                        Morrison Formation

        Ellis Group

                        Chugwater Formation

Generalized vertical relationships of Mesozoic rocks in Montana and vicinity. This is how the rocks would look underground if there were no erosion or areas where mountain-building forces had lifted the rocks. At the bottom of the section, the Chugwater Formation, Ellis Group, Morrison Formation, and Kootenai/Cloverly Formations are essentially flat. The younger terrestrial rocks of the Eagle, Judith River, and Two Medicine Formations all are wedge-shaped, thicker in the west and thinner toward the marine sediments in the east. The wedge shape represents the great volume of sediment deposited in the west as the Rocky Mountains began to rise and erode. Episodes of mountain building caused sea level to change. As a result, the Cretaceous Inland Seaway expanded and contracted, depositing sediments of the Colorado Group, Claggett Formation, and Bearpaw Formation over a wide and shifting area.

If you look at the geological map of Montana, you will notice that the Two Medicine Formation crops out in western Montana along the eastern edge of the Rocky Mountains, and the Hell Creek Formation is found in eastern Montana. The Two Medicine Formation is older than the Hell Creek Formation, even though both formations are exposed at the surface of the ground in their respective regions.

If you drive from west to east on U.S. 2, along the "Highline" from Browning to the North Dakota border, you pass over both Middle and Upper Cretaceous formations. Around Browning the latest Cretaceous St. Mary River Formation is exposed at the ground's surface. Between Browning and Cut Bank, the marine Bearpaw Formation forms the flat plains, and at Cut Bank, the Two Medicine Formation makes the cliffs above Cut Bank Creek. A few miles east of Cut Bank, the road drops over a hill to a valley in which the small town of Ethridge sits. To the north are cliffs of "The Rim," formed by the beach deposits of the Virgelle Sandstone. From Ethridge to my hometown of Shelby and beyond to Lothair, you are in the Colorado Group. It contains two marine formations that were deposited in an intercontinental ocean at the time dinosaurs lived in western Montana. In the 45 miles from Browning to Ethridge, you drop only about 500 feet in actual elevation, but at the same time you pass down through 3,000 feet of strata.

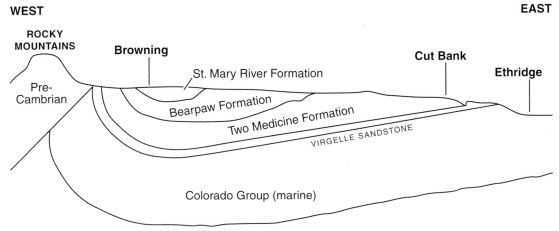

Underground relationship of Cretaceous rocks from Browning to Ethridge along U.S. 2, showing why rocks of different ages—and therefore fossils of different ages—exist at the two locations.

Just east of Lothair, the road rises back onto the Virgelle Sandstone, and then around Chester, the marine Claggett Formation is exposed, although grass mostly covers it. East of Chester, the road climbs toward Joplin and passes into the Judith River Formation. From Joplin to Havre, the road sits atop the Judith River Formation. From Havre to Malta, occasional badlands in the Judith River Formation border the valley of the Milk River. From Malta to nearly Hindsdale, most of the rock belongs to the Claggett Formation. Around Hindsdale, more rocks of the Judith River Formation crop out. From Hindsdale to Brockton is the Bearpaw Formation. At Brockton, the road crosses a small portion of the Hell Creek Formation and then passes into the Paleocene Fort Union Group, which extends all the way to the Canadian border and beyond.

The road passes through all these formations without actually changing much in elevation because the layers within these formations lie tilted rather than flat. Deep, underground movement of molten rock that pushed its way upward and formed mountain ranges such as the Bearpaw Mountains and Little Rockies caused most of the tilting in this area about 30 million years ago. Around Shelby the domelike Sweetgrass Arch bowed the older marine sediments of the Colorado Group. Both east and west of Shelby the strata are tilted, so if you travel away from town, the road actually climbs in time, getting into younger and younger sediments.

In Montana, collectors are most likely to find dinosaur remains in the terrestrial St. Mary River, Two Medicine, Judith River, and Hell Creek Formations. There is little need to look in the marine sedimentary rocks of the Colorado Group or the Claggett and Bearpaw Formations for anything other than marine organisms. Knowing this allows us to better pinpoint the location of potential dinosaur remains. When I was a young boy growing up in Shelby, I was destined to find marine animals because the rocks where I lived were deposited in an ocean and did not contain remains of terrestrial animals such as dinosaurs. But dinosaur remains have been found in the Two Medicine Formation in Cut Bank, only 24 miles west of Shelby.

Dinosaur remains, like many other fossils, are preserved in sediments that either water or wind laid down. Those same depositional processes still operate today. By studying modern river systems, geologists have discovered key features that identify different parts of the stream and river systems. We can determine the environmental setting that existed where the bones or fossil remains came to rest by studying the rocks that contain them, looking specifically for the clues that tell us which part of the river or stream deposited the sediments.

The rivers that flowed through Montana during the age of dinosaurs carried mainly sand and silt, not gravel like we see in many streams and rivers today. Most of the gravel we see in our modern rivers was either pushed south from Canada by glaciers or gouged out of the Rockies during the last ice age. Glaciers did not carve the mountains that existed during Mesozoic time, and they were rounded like the mountains in the eastern United States. Mesozoic rivers carried sand, silt, and even volcanic ash, but very few rocks. The sand has now hardened into sandstone, the silt into mudstone, and the volcanic ash into the clay-rich mineral bentonite, locally called gumbo. The sandstone layers represent the actual stream or river channels. The mudstone strata represent the floodplains of the stream or river. If you find a dinosaur specimen in sandstone, you can deduce that it most likely washed down a stream or river. If you find a bone in mudstone, you will know that a local flood covered it with mud.

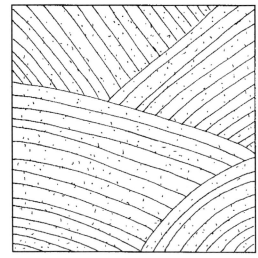

A pattern of crossbedding in sandstone indicates that a river or wind deposited the sediments.

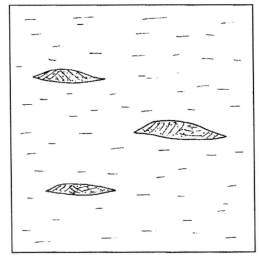

Small lenses of sandstone in mudstone tell us the sediments were laid down by small streams that flooded often.

If you look at the cut bank of a river or a trench through a sand bar or sand dune, you typically see layers that lie at very steep angles. Wind or water laid down these steep beds as they passed over the surface of the dunes and left some sediment behind. When we find a dinosaur in sandstone that has these steep beds, we know that the dinosaur died either in a river or a sand dune. River dunes, or ripples, are usually pretty small, whereas sand dunes can be extremely large. Most of the big red and tan cliffs in Zion National Park in southern Utah are made of huge, ancient sand dunes. Stream and river dunes are commonly only a few feet thick at the most and typically have a layer of gravel at the bottom. Sediments deposited in a small stream might also contain occasional lens-shaped pockets of crossbedded sediments. Floodplain deposits are usually represented by mudstone adjacent to the layers of sandstone. Mudstone seldom contains any observable patterns within its layers. Sediments deposited in ponds and lakes are usually laid down in flat, evenly spaced layers. Swampy deposits most often consist of brown or black siltstones that contain lots of plant debris. Good leaf impressions are common on the tops and bottoms of layers of swamp deposits.

Exceptionally thick geologic formations, such as the Two Medicine Formation, represent the passage of a considerable amount of time. It is important, therefore, that the collector record the

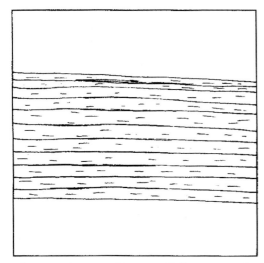

Flat, evenly spaced layers of siltstone and mudstone indicate sediments deposited in ponds or lakes.

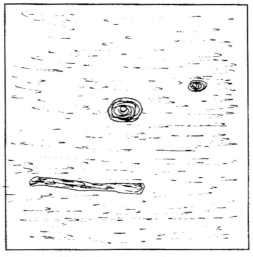

Brown to black siltstones and mudstones with plant debris indicate deposition in a swamp.

approximate stratigraphic level of the fossil find; that is, the point or horizon in the stratigraphic sequence in which the collector found the specimen. Pinpointing the stratigraphic level marks the point in geologic time when sediment buried the specimen. Geologists measure how far certain layers or fossils are either from the top or bottom of the formation. If the formation boundary proves impossible to locate, the collector should at least measure to the top or bottom of the outcrop. That measurement will provide some record of the specimen's relative stratigraphic position.

Taphonomy

Paleontologists are interested not only in the kinds of sediments that yield dinosaur remains but also in how the bones are arranged and their possible association with other fossils. This important aspect of vertebrate paleontology is called taphonomy, the study of the origin and burial of fossil organisms. When a paleontologist finds an accumulation of dinosaur bones, he or she maps and photographs the site and carefully studies it before removing the bones. The researcher records individual bones, plant debris, and invertebrate fossils on the map, and then plots them on graph

Map of a bonebed in the Two Medicine Formation that contains primarily the bones of Einiosaurus procurvicornis.
−From Rogers, 1989

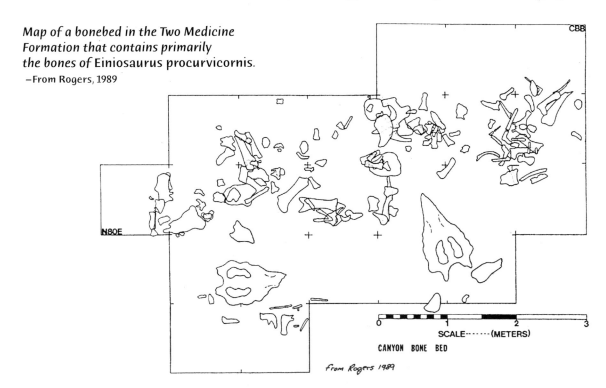

SCALE········(METERS)

CANYON BONE BED

from Rogers 1989

paper to see if there is a pattern to their distribution. If many of the bones line up in a particular pattern, it could mean that flowing water oriented the bones. A jumbled mass of bones with no orientation might record a catastrophic event such as a mud flow or a storm.

After mapping and photographing the bones, the research team excavates them and takes them back to the lab. There they clean them up and study them further, looking for any clues to what happened to the animals just before and just after their deaths. Careful examination might reveal evidence that some of the bones accumulated in episodes at different times than others, or that the bones accumulated continuously over many years. Bones from a number of animals that all died together might imply that a group of animals were living together. Paleontologists carefully examine individual bones for evidence of bite marks, insect borings, trampling, or any other interesting features. Occasionally, they find broken teeth from carnivorous dinosaurs embedded in the bones of plant eaters. Such a discovery is direct evidence that the meat eater ate the plant eater. Careful studies can also reveal pathologies and other aberrations on bones. Pathologies give us clues about dinosaur diseases and wounds. Paleontologists can easily identify pathological bones by their

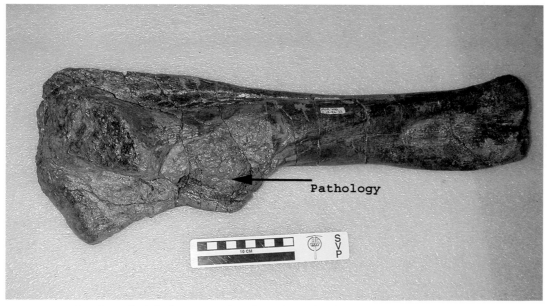

The ulna of Diplodocus. *Notice the enlarged, rough pathology where either disease or traumatic injury caused a swelling on the bone.*

characteristic swollen or abnormal-looking surfaces. We find many dinosaur bones that had been broken and then rehealed while the animals were alive. An *Allosaurus* skeleton from northern Wyoming has numerous pathologies that suggest the animal either had a severe bone disease or had been seriously injured during a fight. All the bones had healed, which suggests that none of the injuries were fatal.

Articulated Skeletons

For most professional and amateur paleontologists, articulated skeletons are the *crème de la crème,* the very best. These are specimens in which the bones remain in the positions they occupied during the animal's life. Having excavated a number of articulated specimens, however, I'm inclined to think that having them in separate pieces allows more thorough study and certainly makes them easier to collect. Articulated skeletons are obviously very important for studies of dinosaur locomotion, stance, and a variety

The articulated skeleton of **Brachylophosaurus** *from the* **Judith River Formation** –Museum of the Rockies, Bruce Selyem photo

of other activities, but unless they are completely dismantled, it is difficult to study individual bones. Because articulated skeletons contain so much valuable information, their excavation is best left to people familiar with dinosaur anatomy. Inexperienced collectors risk destroying or losing important elements.

We typically find articulated skeletons by themselves, commonly in channel sandstones or in sandstones consisting of wind-blown sand that quickly buried the skeleton. Articulated skeletons have been found in all the Cretaceous formations of Montana.

Sometimes we find skeletons that are partially articulated, meaning that some part of the skeleton remains in its original position but water or scavengers have separated other parts. When we find an incomplete skeleton, we typically expand the limits of the excavation to search for some of the missing items. The *Tyrannosaurus rex* skeleton that the Museum of the Rockies excavated in 1990 was partially articulated. Its vertebrae were intact from the first vertebra of the neck, the atlas, to the fourteenth vertebra of the tail. The skull was dislodged to a position above the pelvis. The left leg was articulated, but the right leg had washed behind the skeleton and lodged between the skeleton and skull. The end of the tail, several ribs, an arm, and one foot all were missing. Further excavation revealed a portion of the foot, but nothing more.

Rarely we find several articulated skeletons associated with one another. Upon preparation, the articulated *Brachylophosaurus* skeleton from Malta, Montana, revealed the articulated skeleton of a garfish that evidently died while it was scavenging the insides of the sunken hadrosaur. Apparently, sand rapidly covered both specimens, exquisitely preserving them.

Articulated skeletons commonly have associated skin impressions preserved around many of the bones. Preparation of the specimen must proceed extremely carefully and slowly to avoid destroying these scientifically important impressions. Articulated skeletons may also preserve intricate structures such as the stapes in the ear and thin plates that surround the eyes.

Associated Skeletal Remains
Associated skeletons are those in which the bones have mostly come apart but haven't moved far. We study these skeletons in

The associated ribs of Tyrannosaurus rex called Celeste discovered in the Hell Creek Formation in eastern Montana

great detail, but they give little or no particular information about how the bones fitted together during the animal's life. We find associated skeletons most commonly in floodplain mudstone deposits, where the animals died, decomposed, scattered, and were later covered by sediments during a flood. Most associated skeletons are incomplete, usually missing their skulls, feet, or other parts. Some associations might consist of only a few bones. We always examine each bone carefully to be sure that together they represent one animal. Some close associations represent bones from a number of individuals that simply washed together. Usually we can tell whether the bones belong to one animal if they are the right size for one individual, they are the same color, they have similar erosion patterns, and no duplicate parts are indentifiable.

Bonebeds

In some instances, floodplain strata contain massive accumulations of disarticulated skeletons, or bonebeds. Bonebeds represent any number of individuals. Often these accumulations contain

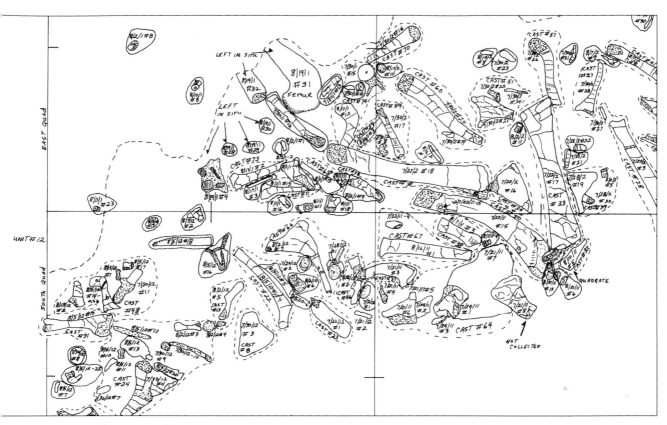

Map of "Camposaur" Quarry showing skeletal elements of several hadrosaurs

bones of mainly one species of dinosaur. A bonebed of *Maiasaura peeblesorum* near Egg Mountain, on Montana's Rocky Mountain Front, covers an area exceeding 1 square kilometer (about 0.4 square mile) and contains thousands of bones representing between 10,000 and 15,000 animals. Taphonomic and geochemical studies of this site indicate that all the animals died together in some kind of catastrophic event, lending evidence to the hypothesis that these dinosaurs lived in a gigantic "herd" prior to their deaths. We commonly find duck-billed dinosaurs and horned dinosaurs in bonebeds, which suggests that all the species of these two major groups of dinosaurs may have traveled in large groups.

Some bonebeds contain the remains of various species that accumulated over long periods of time. They are the result of multiple depositional events. We call these accumulations multispecies bonebeds. Commonly these accumulations are associated with river channel deposits where a number of carcasses washed together in something similar to a logjam. One of the

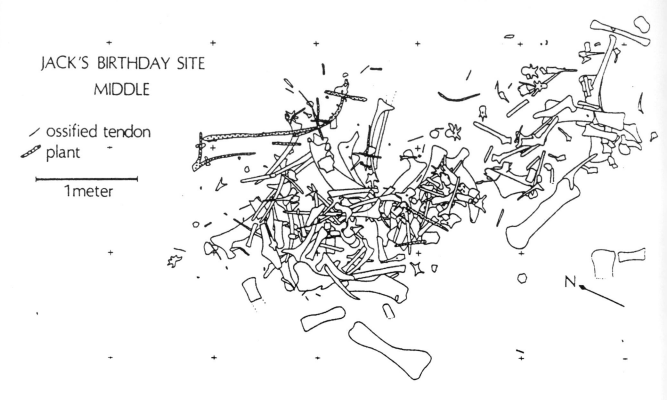

JACK'S BIRTHDAY SITE
MIDDLE

/ ossified tendon
🖉 plant

|⟵ 1 meter ⟶|

N

Map of Jack's Birthday Site, an extensive multispecies bonebed in the Two Medicine Formation –From Varricchio, 1995

most famous multispecies bonebeds is at Dinosaur National Monument in Utah. Of the several accumulations of this type found in Montana, the best is probably Jack's Birthday Site—a site I discovered on lands of the Blackfeet Nation on my birthday in 1988.

Microsites

Microsites are accumulations similar to multispecies bonebeds, but at a much smaller scale. Microsite specimens are small and consist primarily of teeth, small bones, bone fragments, and invertebrate fossils.

We typically find microsite accumulations at the bottoms of sandstone channels, in what geologists call lag deposits. These small teeth and bones moved in the bottoms of streams and rivers as gravel moves in modern rivers. When streams and rivers go around bends or widen out, the water typically slows, and the smaller gravel—or in a microsite's case, bones and teeth—settle to the bottom, where they pile up. Some researchers think that some microsite accumulations of bones and teeth, particularly

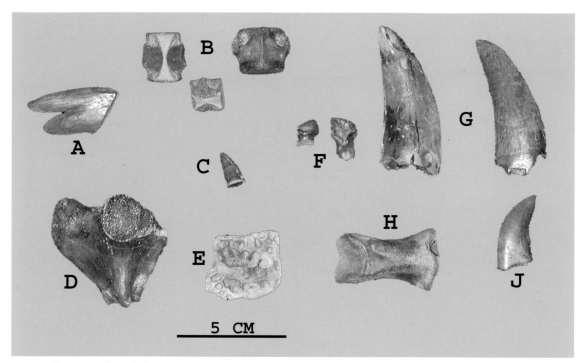

A collection of small bones and teeth recovered from a multispecies microsite in the Judith River Formation: A, portion of turtle carapace; B, *Champsosaurus* vertebral centra; C, crocodile tooth; D, proximal end of turtle femur; E, crocodile scute; F, ankylosaur teeth; G, tyrannosaur teeth; H, ornithomimid phalanx; J, dromaeosaur tooth

those in mudstone deposits, weathered out of a carnivorous animal's dung.

Microsites play an important role in paleontology because they represent many of the different kinds of animals that lived in the rivers, and they include parts of animals that died in or near the rivers. Microsites often yield an abundance of fish, amphibian, and aquatic reptile bones, such as those from crocodiles, turtles, and champsosaurs, along with dinosaur teeth and bone fragments, mammal teeth, and even parts of mollusks.

Isolated Bones

While bones found by themselves can be tough to identify, other than to place them in a major group such as tyrannosaur or hadrosaur, they can be very useful to science. Isolated bones can help researchers determine faunal lists and population structure. They can be useful for histological studies if identified to the genus level. Identifiable bones that are unidentifiable to genus are also good specimens to use for hands-on educational experiences in museums or classrooms.

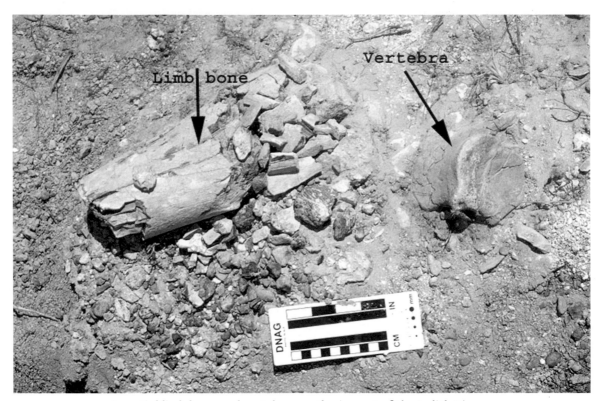

A partial limb bone and vertebra weathering out of the Judith River Formation –Museum of the Rockies photo

Some rare dinosaurs are known only from isolated bones, so it is important to pick up and record anything out of the ordinary.

Preservation of Dinosaur Fossils

The remains of dinosaurs or other extinct organisms consist of any trace of the animals' existence. Evidence for the existence of dinosaurs includes their skeletons, impressions of skin and soft tissue, footprints, eggs and nests, gizzard stones, and even remains of their feces or dung. Fossil dung is called coprolite.

Skeletal Remains

The skeletons of all animals consist primarily of the mineral calcium phosphate, which is very stable and generally resists degradation. After an animal has died and rotted, and sediments have buried its skeleton, groundwater commonly fills the hollow spaces and pores within the bones. Groundwater contains a lot of minerals, and these minerals precipitate, or crystallize, within the bones in much the same way that minerals are deposited in water pipes. This process of filling the cavities within bones, called permineralization, is why fossil bones are especially heavy. The original calcium phosphate might still be present, but the water-deposited minerals have been added to the bone. Sometimes another mineral partially or completely replaces the calcium phosphate in a process called replacement. Replacement occurs molecule by molecule and takes a long time. When minerals replace wood, we say the wood is permineralized or petrified.

When sediments cover bones and subject them to permineralization or replacement, the bones commonly take on different colors. The colors usually depend on the mineral that is incorporated within the bone. Dinosaur bones that we find in some formations in Montana are black, whereas those we find in other formations are brown or even reddish. Once the bones weather out of the sediment, the sun may bleach them to white or tan. Even bone that is not permineralized or replaced can change colors after long periods of burial. Buffalo bones only about one hundred years old are commonly brown from taking up minerals while buried.

Most dinosaur bones from the Upper Cretaceous sediments of Montana are extremely well preserved, having been permineralized but not entirely replaced. In some rare instances, they have even yielded biomolecules such as proteins. Looking through a microscope at a cross section or thin section—paper-thin slices of bone mounted on glass slides—it is possible to see cellular structures in most dinosaur bones. Studies of these structures are revealing evidence of dinosaur growth rates, age at death, and even indications of their metabolism.

Soft Tissue

Dinosaur skin was a soft tissue, and preserved dinosaur skin has never been reported. However, impressions of dinosaur skin exist that reveal the actual surface texture of the animal's exterior. The impressions can even be three-dimensional. The impressions were made when the animal laid down or died and pressed its skin into the clay or silt. When the tissue rotted, the impression remained, and other sediment later filled in around it. Dinosaur skin is never preserved on the surface of bone but rather in the sediments that surround the bone.

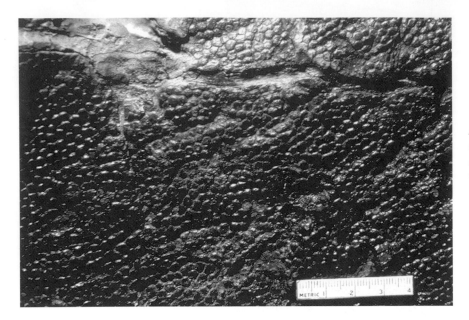

A section of hadrosaur skin impression from the Judith River Formation —Museum of the Rockies photo

Skin impressions are best preserved in extremely fine-grained sediments such as limestone, siltstone, and clay. Coarse-grained sandstones do not preserve these structures. Skin impressions most typically are associated with articulated skeletons, but on rare occurrences collectors find small patches of skin impressions in bonebeds. Skin impressions associated with a skeleton require extremely careful excavation and preparation because the impressions have no actual substance.

Skin impressions are known for a few duck-bills, horned dinosaurs, and some theropods and sauropods. Dinosaur skin impressions show that they had bumpy or pebbly skin and that they were not scaly. The skin impressions of duck-bill dinosaurs show that raised, round bumps covered their bodies. Some dinosaurs had skin frills that extended down their necks, backs, and tails, as many lizards do. Some of these frills were segmented and others were continuous.

A couple of duck-bill dinosaur skeletons preserve an abundance of skin impressions that show that the skin had dried up and pulled tight over the skeleton. These specimens, which some call mummified dinosaurs, are not actually mummies because there is no preserved skin. The "skin" is only an impression in the sediment.

Footprints

Dinosaurs also left evidence of their existence in the form of footprints. In many regions of the world, long trackways of dinosaur footprints show how various species walked or ran. Some trackways have even yielded behavioral information, such as showing that two animals walked side by side.

The preservation of footprints can take place in two ways: by preservation of the original indentation, called the footprint mold, or by preservation of a filled footprint mold, called a cast. The long trackways of dinosaur footprints, such as those in Texas or Colorado, that you might see in photographs are footprint molds. They are the original footprints that the animals made as they walked across mudflats, probably near rivers or lakes. A few days or weeks after the animals passed through, sand filled the footprint depressions. Continued deposition of sand and mud buried the footprints until much later, when they were once again brought to the surface and weathered. Being softer than the mudstone of the former mudflat, the sandstone filling weathered out and left preserved footprints.

Footprint casts form in the same way as molds, but they weather differently. The dinosaurs leave their footprints in mud, and the impressions later fill with sand. In this case, though, the mudstone is softer than the sandstone. So the mudstone in which

*Cast of a theropod dinosaur footprint from
the Morrison Formation of Utah* –C. C. Horner photo

the dinosaurs walked weathers away and leaves behind the harder sandstone casts that had filled the footprints. Footprint casts are harder to find than molds because casts are not indentations, but rather foot-shaped rocks.

Unfortunately, dinosaur footprints are rare in Montana. Most of those that have been found are casts rather than molds. Footprint casts are known from nearly all the Cretaceous formations of Montana, and a few, rare footprint molds are known from the Jurassic Morrison Formation. Footprint molds are best left in place and copied using rubber molding compounds. Removing footprints, especially if they are part of a longer trackway, is destructive and not considered good science.

Eggs and Nests

Fossil dinosaur eggs and nests tell about the social behaviors of dinosaurs. Dinosaur eggs vary from about the size of a baseball to that of a soccer ball. They commonly have bumps or ridges on their surface. Most dinosaur eggs and eggshell fragments are

A clutch of Maiasaura eggs
from the Two Medicine Formation
—Museum of the Rockies,
Bruce Selyem photo

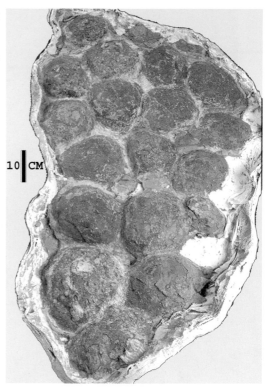

A clutch of lambeosaurine eggs
from the Judith River Formation
—Museum of the Rockies,
Bruce Selyem photo

associated with nesting grounds. The eggs of most dinosaur species are unknown.

We can only assign dinosaur eggs to particular dinosaur groups based on comparative features or embryonic remains within the eggs. Hadrosaur eggs are spherical, and theropod eggs are elongated and typically asymmetrical—blunt on one end and pointed on the other, like a bird's egg. The egg's shell proves much more useful than its shape in identifying dinosaur eggs. In fact, some rocks that look like whole or partial eggs are not eggs but concretions (see chapter 6, pseudofossils). Dinosaur eggshell from Montana is generally no thicker than a couple of millimeters ($\frac{1}{16}$ inch). Dinosaur eggs were hard-shelled like the eggs of birds, not leathery like modern lizard eggs. Dinosaur eggshell fragments, though uncommon, have been found in nearly all Cretaceous sedimentary rocks in Montana.

Nest structures such as rim boundaries have been found in Montana, but they are extremely rare. The little, carnivorous

A Troodon nest showing the rim that surrounded the egg clutch. The white area is a plaster jacket covering the eggs.
—Dave Varricchio photo

A clutch of Troodon *eggs from the Two Medicine Formation* —Museum of the Rockies, Bruce Selyem photo

Portion of a coprolite of Maiasaura *from the Two Medicine Formation*

dinosaur *Troodon* created a sediment rim around its egg clutch that was about 12 to 15 centimeters (5 to 6 inches) high, and the duck-bill *Maiasaura* dug a nearly 2-meter (6.5-foot) diameter pit in a mound of dirt. Some dinosaurs appear to have simply scraped away a little dirt, laid their eggs, and then covered the eggs with sediment. Ornithischian dinosaurs and sauropods probably covered their eggs with sediments, whereas many of the theropods may have brooded their eggs by sitting on them, as most living birds do.

Coprolites and Gizzard Stones

Fossil dung, or coprolite, allows scientists to determine what some dinosaur species actually ate. In some cases, paleontologists can attribute coprolite specimens to a specific group of dinosaurs. The duck-bill dinosaur *Maiasaura* apparently ate mainly conifer leaves, and *Tyrannosaurus rex* ate duck-bill and horned dinosaurs. Burrows in the coprolites of *Maiasaura* reveal that dung beetles fed on the dinosaur dung. Coprolites of plant-eating dinosaurs are usually found as masses of mashed plant debris, while coprolites from carnivores contain bone fragments that stomach acids have partially dissolved. Very few of the rocks sold as dinosaur coprolites are the real thing—if you are unsure, get the opinion of a dinosaur researcher.

Gizzard stones are rocks that reptiles and birds ingest to grind food in their crops or stomachs or that an animal such as a plesiosaur uses as ballast to keep itself underwater. Some scientists suggest that some extremely shiny, rounded rocks in the Cloverly Formation of Montana were dinosaur gizzard stones. These rocks are highly polished as though they spent time in a rock tumbler. They range from thumb size to fist size. Whether or not these rocks are actually gizzard stones, they certainly did not form from ordinary geological processes. Similar rocks are occasionally found in the Morrison Formation, and researchers reportedly found one group in the neck region of the sauropod *Seismosaurus* from New Mexico. This association suggests that at least the sauropod dinosaurs used gizzard stones, although other researchers now challenge that hypothesis.

History of Dinosaur Collecting in Montana

As a Montana native, I get a real kick out of knowing that the first dinosaur remains found in the Western Hemisphere came from a part of the Nebraska Territory that is now central Montana. A geologist named Ferdinand Vandiveer Hayden, working for the United States Geological Survey, spent a great deal of time exploring the Nebraska Territory in Sioux country. The Native Americans left him alone, believing he was a harmless madman. They called him "he who picks up stones while running." In 1855, while exploring the upper reaches of the Missouri River, Hayden came across an area of badlands near the mouth of the Judith River. There, he picked up a small collection of teeth, bones, and shells. Upon his return east, he took them to Philadelphia, where Professor Joseph Leidy, an anatomist at the University of Pennsylvania, studied them.

Leidy recognized the teeth as being different from any other teeth ever found in North America, but somewhat similar to the teeth known from England that scientists had previously referred to dinosaurs. Leidy gave new dinosaur names to the small collection of teeth: *Trachodon* (rough tooth) *mirabilis*, *Deinodon* (terrible tooth) *horridus,* and *Troodon* (wound tooth) *formosus.* So little was known about fossilized animals in those days that it was customary to give new taxonomic names even to isolated elements such as individual bones and teeth. Almost everything that was found, no matter how fragmentary, was probably new to science. In the middle of the twentieth century, this practice was discouraged, but it was common during the late 1800s.

Two years after Leidy described the Montana dinosaur remains, quarry workers found a much better dinosaur skeleton in

New Jersey that Leidy named *Hadrosaurus,* or "bulky reptile." *Hadrosaurus* was a duck-bill dinosaur. It was the most complete skeleton of a dinosaur known at that time and the first evidence to conclusively show that some dinosaurs were bipedal. Teeth associated with *Hadrosaurus* were similar to the Montana *Trachodon* teeth, so duck-bill dinosaurs were known for a long time as trachodonts. We now refer to them as hadrosaurs because *Trachodon* is not accepted as a valid taxon. The name *Trachodon* is based on a single tooth, and at this point, we cannot identify isolated hadrosaur teeth to the genus level. Certain characteristics of the teeth, however, allow tentative identification to either the flat-headed hadrosaurine or crested lambeosaurine duck-bills.

The second paleontological expedition to what is now Montana took place in 1876. The team included three of America's most prolific collectors: Edward Drinker Cope, who was a student of Leidy's from Philadelphia, Charles H. Sternberg from Kansas, and J. C. Isaac from Wyoming. Cope and his assistants Sternberg and Isaac traveled 600 miles by stagecoach from the narrow-gauge railhead near Franklin, Idaho, to Helena, where the three men secured horses and a wagon to take east to the Fort Benton area. While the three were in Helena, the local newspaper reported the story of the massacre at the Little Big Horn, in which General Custer met his fate. The news of the massacre didn't sway Cope, who instead of being fearful thought it was the perfect time to travel to the Judith River and collect dinosaur fossils. Cope believed the Sioux, under Chief Sitting Bull, would be otherwise occupied following their victory. The Crow he believed to be of no threat.

The men left Helena and traveled north to Great Falls and then to Fort Benton, where they hired a cook and a scout. They drove their buckboard to an area near the mouth of the Judith River near Fort Claggett. They continued on to set up camp at Dog Creek, across the Missouri from the lodges of two thousand Crow who were preparing for their annual buffalo hunt. At that time, all the land north of the Missouri River, the Marias River, and Birch Creek—from the Continental Divide east nearly to what is now the Montana–North Dakota border—was the reservation land of the Gros Ventre, Piegan, Blackfeet, and River Crow people. On a couple of occasions, Cope and his crew invited some of the Crow chiefs to evening meals and let them know that he and his fossil hunters would be in the area. With the politics taken care of, Cope and his crew set out to explore the badlands along the Missouri River from Judith Landing east to Dog Creek—the area now known as the Missouri River Breaks. That area has always been very rugged, and no one has inhabited it for more than a century. But during the days of steamboat travel, a number of people inhabited it. There was a stockade at Claggett, opposite the mouth of Judith River, a steamboat landing at Cow Island, and numerous outposts and woodcamps all along the river. Woodhawks, who gathered wood for the paddleboats to burn for their steam power, inhabited the woodcamps. The Cow Island landing was as far west as the boats could travel during the fall, when the river was low.

Cope and his team collected without incident. They found hundreds of bones and teeth, many of which Cope would later make the basis of new species of dinosaurs. We now consider nearly all these dinosaur species invalid because not enough of any of the specimens exists to show a particular uniqueness. At the end of the season, on October 15, Cope loaded his paleontological treasures onto *Josephine,* the last paddleboat of the year at Cow Island, and made the journey back to Omaha and finally Philadelphia. The trip to Omaha took about a month.

Sternberg, who would become a major paleontological figure in Alberta, would return to Montana in 1914, revisiting the sites where he and Cope originally collected in 1876.

Another nineteenth-century collector was J. B. Hatcher, a paleontologist from the United States Geological Survey. In 1888, he collected a couple of hadrosaur jaw specimens and a partial skull of a ceratopsian for Yale University. O. C. Marsh, vertebrate paleontologist for the Geological Survey at the time and also a professor at Yale University whose family founded its Peabody Museum, named the hadrosaurs *Hadrosaurus paucidens* and *Hadrosaurus breviceps*, and the ceratopsian *Ceratops montanus*. Neither hadrosaur specimen is identifiable beyond being that of a duck-bill dinosaur, so both are considered invalid today. *Ceratops montanus* is also very fragmentary, but it may one day prove to be valid. It shares some characteristics with *Avaceratops*. Hatcher is known for discovering the first remains of *Triceratops*, found near Lusk, Wyoming, and important Cretaceous mammals. J. B. Hatcher collected his dinosaur fossils from the same areas Cope had visited the year before.

Montana's first graduate student, Earl Douglass, discovered the first nearly complete dinosaur skeleton reported from Montana. Douglass was majoring in geology and studying mammal paleontology at the University of Montana in Missoula in the late 1800s. In 1899 and 1900, he and his assistant, A. C. Silberling, a native of Harlowton, discovered four dinosaur skeletons, two in the vicinity of Big Lake near Rapelje and the other south of Harlowton. Three were duck-bills, and one was an armored dinosaur. One of the hadrosaurs was preserved with skin impressions and is nearly complete. One hadrosaur specimen, taken to Missoula, was apparently lost. Douglass shipped the other specimens to Princeton University, where he completed his graduate studies. Having studied the Princeton specimens, I believe one of the hadrosaurs to be a species of *Gryposaurus,* and the other may be a lambeosaur, but I might be wrong about both of them. The armored dinosaur is a nodosaur. Later in his life, Earl Douglass

Site of the first Tyrannosaurus rex *skeleton, which Barnum Brown found in 1903 on Hell Creek. The site is on the side of the hill behind the tent.*
–American Museum of Natural History photograph

Barnum Brown's Tyrannosaurus rex *skeleton on Hell Creek during its excavation* —American Museum of Natural History photograph

discovered what is now called Dinosaur National Monument on the border of Utah and Colorado.

In 1903, a collector from the American Museum of Natural History named Barnum Brown began an extensive exploration of Montana and Alberta that continued until 1939. One of his first discoveries was the partial skeleton of a dinosaur that Henry Fairfield Osborn would name *Tyrannosaurus rex*. Brown found the skeleton in the Hell Creek Formation of eastern Montana. It was the first dinosaur specimen Brown found after arriving at his campsite. He later found a second specimen of *T. rex* and a number of specimens of *Triceratops* and *Edmontosaurus,* then called *Trachodon*. Brown spent only a few years in the Hell Creek Formation before he moved farther west. Brown apparently didn't spend much time in the Missouri River area where Hayden, Cope, and Hatcher originally collected, but instead searched other Mesozoic formations around the state.

In 1916, Brown turned his attention to the Two Medicine and St. Mary River Formations in northwestern Montana. In the Two Medicine Formation, Brown collected numerous skeletons of duck-bill dinosaurs, one of which was a new species. Brown never got around to publishing his description of the skeleton, so I wrote it up a few years ago, naming it *Gryposaurus latidens*. In 1987, the man who originally found that dinosaur told me the story of its discovery. I had stopped at a rancher's house to ask permission to set up a camp on the Two Medicine River and to collect in the area. The ranch owner, an old, distinguished fellow named Tom Harwood, gave me a long, hard stare before he explained that I could camp on the river but that I would not be able to camp where his last paleontology visitors had camped. Seems as though the flood of 1964 had washed away the campsite. I was shocked to hear that other collectors visited the area during the sixties and asked him what institution they were from. Tom looked up at the ceiling, thought for a couple minutes, and then said, "Wasn't the sixties, earlier. A professor type from New York." "New York? Do you mean Barnum Brown?" I asked skeptically. "Yup," he replied. "Brown, that's it. Professor Brown!" "That was 1916," I said. Tom grabbed his chin, deep in thought, and replied, "Yup, that sounds about right. Like I said, a few years back." Tom was born in 1900 and was sixteen when Brown visited his family's ranch.

Tom said that he had found the original skeleton of the big duck-bill and that Brown had given him one hundred dollars for the jaw and the location of the specimen. Tom took us to the site, which was obviously the site of an old excavation. We found a rusted awl and a few tin cans just below the site. Tom later explained that he never actually saw Brown at the excavation site, but rather his crew of young students from New York. Tom said that Brown spent most of his time in the town of Cut Bank, 15 or so miles away. Apparently Brown came out only to decide what was to be removed and sent back to New York. Unfortunately, Brown didn't take very good notes, and it is extremely difficult to relocate most of his localities.

After only a couple of years working in the Two Medicine Formation, Brown turned his exploration to Alberta. He returned to Montana a few years later to work in the Lower Cretaceous

Charles Gilmore and George Sternberg excavating a dinosaur from the Two Medicine Formation —Smithsonian Institution photograph

Cloverly Formation and then the Upper Jurassic Morrison Formation. He collected a lot of dinosaurs but never got around to publishing descriptions of any of the specimens from either the Jurassic or Lower Cretaceous formations. He did, however, cover a great deal of ground, visiting nearly every area now known to produce dinosaurs.

Gatz Hortzberg, a writer friend of mine, once took me to the top of a mountain near his home near Big Timber to show me a site that Barnum Brown apparently visited in 1939. Brown had obviously not collected anything, but there were tell-tale signs that the site had been examined. The Jurassic skeletons were encased in hard sandstone at the bottom of a 30-foot-high sandstone cliff. An interesting site, but it is unlikely ever to be excavated without heavy equipment. From the few bones weathering out it appears to be a multispecies bonebed, similar to most sites in the Morrison Formation of Montana.

Brown also visited and collected from a site in the Kootenai Formation near Great Falls. There he excavated a very badly preserved

Gilmore and Sternberg quarry containing juvenile hadrosaurs from the Two Medicine Formation
—Smithsonian Institution photograph

skeleton that has yet to be identified. Although interesting from the perspective that no other dinosaur skeletons have been found in the Kootenai Formation in this area, the specimen's poor preservation means that it provides little scientific data. Dr. Jack McIntosh, who studies sauropods, has suggested that the specimen might be an armored dinosaur, and Dr. Walter Coombs, who studies armored dinosaurs, thinks it is probably a sauropod.

During the early 1900s, Charles Gilmore, the Curator of Fossil Reptiles at the United States National Museum (now the Smithsonian Institution), made three trips to Montana to collect dinosaur remains from the Upper Cretaceous Two Medicine Formation on the Blackfeet Reservation. Eugene Stebinger, a geologist for the United States Geological Survey, had made initial discoveries in this area. Stebinger had located the partial skeleton of a large hadrosaur in badlands along the Milk River in 1912. The following year, Gilmore and his assistant, J. F. Strayrer, both working under the sponsorship of the United States Geological Survey, collected Stebinger's hadrosaur and a number of other dinosaur remains, including the first specimens of *Brachyceratops montanensis*. Gilmore and Strayrer discovered a bonebed containing the disarticulated parts of five equal-size specimens of *Brachyceratops*. In the same publication where Gilmore described *Brachyceratops* and a few other remains, Stebinger described the geology, mentioning an extensive stratum that contained abundant shell fragments he believed mainly represented the freshwater clam *Unio*. These shell fragments would be later identified as dinosaur eggshell.

Gilmore returned to the Two Medicine Formation in 1928, and for a last time in 1935. Charles H. Sternberg's son George F. Sternberg and George Pearce accompanied Gilmore during his 1928 visit. The party discovered several new dinosaur specimens, including a bonebed of small hadrosaurs, the skull and skeleton of the armored dinosaur Gilmore would name *Palaeoscincus rugosidens* (now called *Edmontonia rugosidens*), and the frill of the horned dinosaur *Styracosaurus ovatus*. The 1935 visit yielded the fragmentary remains of two specimens of an animal closely related to *Leptoceratops* and what Gilmore thought at the time to be an adult specimen of *Brachyceratops*. The Two Medicine Formation would not be searched again for forty years.

Through the 1940s and 1950s, professional paleontologists made no significant dinosaur collections from Montana. But in the early 1960s, Marshall Lambert, a schoolteacher in Ekalaka, Montana, began a collection effort that would become the collections and displays of the Carter County Museum, the first dinosaur museum in Montana. Marshall traveled back east to Princeton, Harvard, and the Smithsonian Institution to learn curation and preparation techniques from specialists. He convinced the geology department at Princeton to donate display cases for his museum. At Harvard, he learned how to mount a dinosaur skeleton. You can see the results of Marshall's work at the Carter County Museum in Ekalaka. The museum contains a mounted skeleton of the duck-bill dinosaur *Anatotitan,* the skull of *Triceratops,* parts of *Tyrannosaurus rex,* and numerous other important specimens. Marshall worked with paleontologists from other institutions, aiding their research projects by showing them important localities. Marshall's collecting efforts span nearly half a century, from the early 1960s into the 1990s.

John Ostrom and Jack Horner at the Deinonychus quarry that John discovered in 1964 in the Cloverly Formation —C. C. Horner photo

Following Barnum Brown and Charles Gilmore, the first professional paleontologist to take an interest in Montana dinosaurs was John Ostrom of Yale University. John and his students, among them Bob Bakker, began exploration in 1964 in the Lower Cretaceous Cloverly Formation of south-central Montana. Ostrom struck it rich that year with the discovery of several well-preserved dinosaur skeletons, including a bonebed containing the remains of the small meat-eating dinosaur he would later name *Deinonychus antirrhopus*. Ostrom's studies of this dinosaur, together with studies of *Archaeopteryx*, initiated what Bob Bakker would later call the "dinosaur renaissance."

From the original discoveries of dinosaur remains in England during the mid-1800s to the 1964 discovery of *Deinonychus*, most paleontologists held that dinosaurs were little more than overgrown lizards that plodded around, dragging their tails and waiting to go extinct. Dinosaurs were considered big, cold-blooded, stupid reptiles. Richard Owen, the man who originally coined the name *dinosaur* in 1842, thought dinosaurs were more similar in some respects to birds and mammals because their hipbones revealed they stood upright, but the overwhelming opinion was that they

Harley Garbani checking out an old sheep wagon
—Bill Clemens photo

were simply overgrown lizards. Thomas Henry Huxley, a great supporter of Darwin, and Edward Drinker Cope, an Owen follower, also believed that dinosaurs were birdlike, but it was Ostrom who discovered derived characteristics that birds share with theropod dinosaurs, showing they had a common ancestor within the Dinosauria. In a separate study of dinosaur trackways and skeletons, Ostrom also demonstrated that dinosaurs did not drag their tails but instead walked with their bodies and tails parallel to the ground. John Ostrom pioneered the work that initiated the move away from thinking about dinosaurs as big, stupid reptiles to envisioning them as more agile and intelligent animals, as in the movie *Jurassic Park*.

Following Ostrom's work in central Montana, many parties from various institutions began collecting efforts throughout Montana during the 1970s and 1980s. Some searched the Mesozoic sediments for primitive mammals but still collected a few dinosaur remains. Others focused their interest on dinosaurs but collected other creatures as well. Important collecting groups included teams

Princeton University crew collecting Maiasaura *remains from the Two Medicine Formation. Jack Horner is standing to the left, Jill Peterson is sitting on the left, Bob Makela is standing with his back to camera, Wayne Cancro is standing behind Bob, and Jason Horner is sitting beneath Wayne.* —Museum of the Rockies photo

from Harvard University led by Farish Jenkins, the University of California at Berkeley teams led by Bill Clemens, the University of Pennsylvania teams led by Peter Dodson, and the Princeton University and Montana State University (Museum of the Rockies) teams led by my good friend Bob Makela and me. The Harvard group worked primarily in the Lower Cretaceous Cloverly Formation. Besides discovering some very important mammal specimens, they also found the first skeleton of the little ornithiscian *Zephyrosaurus schaffi*. The Berkeley group concentrated on the Upper Cretaceous Hell Creek and Judith River Formations, also looking for mammals, but they found important dinosaur remains as well. Harley Garbani, who helped lead field crews for the Los Angeles County Museum and Berkeley's Museum of Paleontology, is credited with having discovered numerous *Triceratops* specimens, edmontosaurs, and also a skeleton of *Tyrannosaurus rex*. The Pennsylvania team worked the Judith River Formation in central Montana, excavating a new horned dinosaur that Peter Dodson named *Avaceratops lammersi*.

From the late 1970s through the present time, my teams have also made some important discoveries. In 1978, Bob Makela and I were fortunate to get a tip from Bill Clemens about a dinosaur that needed identification in the town of Bynum, northwest of Great Falls. Bob and I went there and identified the dinosaur as a hadrosaur. Then Marion Brandvold, the owner of the local rock shop, showed us some small bones that I identified as those of baby hadrosaurs. Marion turned over the fossils she had to Princeton University after we explained their scientific importance. Bob and I later excavated a nest full of baby skeletons from the site Marion had discovered. The experience began an important chapter in dinosaur paleontology.

Further exploration and excavations from the Two Medicine Formation in the area where Marion had found the babies revealed many exciting finds. Laurie Trexler originally located a new duck-bill species that we named *Maiasaura peeblesorum*. We found a new ornithiscian closely related to *Zephyrosaurus* that my colleague and friend Dave Weishampel and I named *Orodromeus makelai* in honor of Bob Makela, who died in an automobile accident in 1987. At this site, we found the first clutches of dinosaur eggs from the Western Hemisphere and, eventually, the first dinosaur embryos ever found in the world. We found egg clutches and embryos of both *Maiasaura* and the little meat-eating dinosaur *Troodon formosus*, which Joseph Leidy originally named. The eggs, babies, and other discoveries led to new thinking about dinosaurs that included ideas about their social behavior and how some of them may have fed their young like many birds do.

In the late 1980s and throughout the 1990s, I had other teams working in many different areas of Montana. In the Two Medicine Formation on the Blackfeet Reservation, where both Charles Gilmore and Barnum Brown made important discoveries, we found more nesting sites and a number of new dinosaurs. Among them were three duck-bill species and two horned dinosaurs. Two of the duck-bills I named *Prosaurolophus blackfeetensis* and *Gryposaurus latidens,* and one that Phil Currie, of the Royal Tyrrell Museum of Palaeontology in Drumheller, Alberta, and I named *Hypacrosaurus stebingeri.* Then-graduate-student Scott Sampson named the horned dinosaurs *Einiosaurus procurvicornis* and *Achelousaurus horneri. Achelousaurus* is the weirdest dinosaur I think I've ever seen—

The Wankel T. rex *during excavation and after rock had been removed from overtop* —Museum of the Rockies photo

maybe that's why Scott named it after me. *Achelousaurus* is a ceratopsian, meaning that it belongs to the group of dinosaurs that have horns like *Triceratops* and *Einiosaurus*. Instead of horns, though, *Achelousaurus* has areas of rough pits and ridges that resemble diseased bone. Our teams found many other dinosaurs on the Blackfeet Nation, several of which still remain undescribed and unnamed. The process of studying and naming them takes time and patience and must be done with care and precision.

In 1990, the Museum of the Rockies got an opportunity to excavate one of the most complete skeletons of *Tyrannosaurus rex*. Amateur paleontologist Kathy Wankel found the specimen on Fort Peck Reservoir. When we excavated the specimen, it turned out to be the first *T. rex* specimen discovered with preserved arms. It was also one of the largest skeletons ever found, but another *T. rex* skeleton found in South Dakota the same year and affectionately called "Sue" soon overshadowed it.

Other interesting and important discoveries in the early 1990s included two mass graveyards of juvenile specimens of *Diplodocus* from Jurassic rocks of the Morrison Formation. I was fortunate

Ken Olson with Torosaurus skull found in Hell Creek Formation
—Ken Olson photo

to discover the first one on a ranch Ted Turner owned, and amateur collector Sonja Pedilla found the second one on public land that the Bureau of Land Management manages in south-central Montana. Diane Gabriel and Vicki Clouse made another very important discovery in the Judith River Formation of north-central Montana. The two women, both graduate students of mine, discovered extensive nesting grounds of crested duck-bill dinosaurs. They found clutches of eggs and even some eggs with embryos. Vicki continues to work these important sites. Then-graduate student Dave Varricchio excavated Jack's Birthday Site in the Two Medicine Formation, revealing an extensive multispecies bonebed interpreted as carcasses washed up on the shores of a shallow lake. A group of *Troodon*, including two adults and a juvenile, suggested some sort of social group.

In the middle to late 1990s, amateur paleontologists made several important discoveries. Ken Olson of White Sulfur Springs, formerly of Lewistown, discovered the sacrum of a *Triceratops* that a *T. rex* had clearly chewed on. Ken also excavated the largest known skull of *Torosaurus*, measuring 2.75 meters (9 feet) long,

Excavation of associated Tyrannosaurus rex skeleton at the G-rex site during the 2000 field season. Bob Harmon is hammering and Nels Peterson is bending over while Kim Wendel, Joe Beaman, and Yvonne Williams look on.

and is responsible for the discovery of a second *Torosaurus* skull that our museum crew excavated. Nate Murphy, from Malta, discovered a beautifully preserved intact skeleton of the hadrosaur *Brachylophosaurus*, and Harley Garbani, in retirement, discovered the skull of a baby *Triceratops*. More *T. rex* specimens turned up in 2000. One was found on land near Fort Peck Dam, and joint Museum of the Rockies/Berkeley field crews found five skeletons during the 2000 field season. We gave each a letter designation corresponding to its discoverer's first name: B-rex, discovered by crew chief Bob Harmon; L-rex, discovered by camp manager Larry Boychuk; C-rex, discovered by my wife, Celeste Horner; G-rex, discovered by Berkeley graduate student Greg Wilson; and J-rex, discovered by me.

Amateur and professional paleontologists excavate new and important dinosaur specimens every year in Montana, and every so often someone finds a new species. I believe we have found less than 1 percent of the different species of dinosaurs that once lived in this region. There is so much more to discover and so much more dinosaur collecting history to be made!

Mesozoic History of Montana and Vicinity

THE STRATIGRAPHIC RECORD OF ROCK preserved throughout our region records Montana's prehistoric past. We can interpret the rocks by studying certain geological features and the fossils within the rocks, and then we can make a good guess about what kind of changes took place throughout the 165 million years of Mesozoic time.

During the entire Mesozoic era, from 230 to 64.5 million years ago, the world was ice free and temperate. Crocodilians and other cold-blooded animals lived as far north as northern Alberta. Over the 165 million years, North America broke away from the supercontinent of Pangea and drifted from near the equator to its present position. Studies of ancient coastlines by Alan Smith and his colleagues at Cambridge University allow us to track the geographic changes that took place as the continents separated and moved throughout Mesozoic time.

Through Mesozoic time, major changes in sea level at times divided North America into two subcontinents. Shallow seas alternately covered and uncovered the geographic area that is now Montana. Between these marine episodes, coastal plains provided habitat for dinosaurs. The geographic changes created environmental changes that drove the evolution of the dinosaurs. From late Jurassic through Cretaceous time, four major assemblages of dinosaurs inhabited the Mesozoic coastal plains that are now Montana.

TRIASSIC TIME
Scythian Stage
246 to 241 million years ago

Near the beginning of Triassic time, 245 million years ago, the region that is now Montana sat at about 30 degrees north

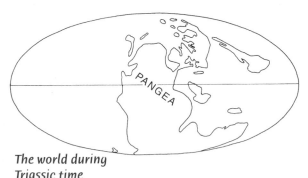

The world during Triassic time

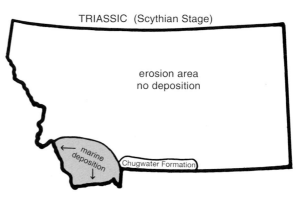

erosion area
no deposition

marine deposition

Chugwater Formation

The depositional features in Montana during the Scythian Stage of the Triassic Period. Triassic sediments lie only in areas of deposition; areas of erosion were not preserved. –From Horner, 1989

latitude on the supercontinent of Pangea. No mountain ranges rumpled western North America, and a shallow seaway reached from the Pacific Ocean over what is now the entire west coast and northward across northern Canada to the Arctic. A small arm, or embayment, of the shallow sea extended into southwestern Montana. Shells of clams and squidlike cephalopods that we occasionally find in these marine sediments tell us that this was a shallow sea.

An extensive desert with sand dunes and small streams covered south-central Montana. The preserved sedimentary formation that contains the sand dunes and stream sediments is called the Chugwater Formation. It is bright orangish red and was deposited about 235 million years ago. You can best see outcrops of this formation east of Bridger, Montana, and on the Wild Horse Range near the Montana-Wyoming border north of Lovell, Wyoming. The remainder of Montana apparently was low in elevation and experienced little or no deposition of sediments. As a result, no record of either sediments or life-forms was preserved. No plant or animal fossils have been discovered in the Triassic Chugwater Formation of Montana, although a few nondinosaurian remains are known from the formation in northern Wyoming. More exploration in this region should reveal vertebrate remains, possibly including some important dinosaurian ancestors.

JURASSIC TIME
Kimmeridgian Stage
150 to 146 million years ago

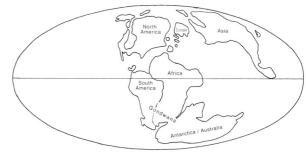

The world during Jurassic time

In the early part of late Jurassic time, parts of Pangea began breaking up. North America pulled apart from South America and Africa and drifted north. A sea submerged much of western North America. An embayment similar to the one that existed during Triassic time extended eastward and covered most of Montana. Only a small island rose above the

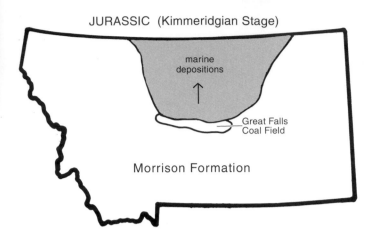

JURASSIC (Kimmeridgian Stage)

marine depositions

Great Falls Coal Field

Morrison Formation

The depositional features in Montana during the Kimmeridgian Stage of Jurassic time

sea in what is now southwestern Montana. Pelecypods, cephalopods, crinoids (also known as sea lilies), and reptilian ichthyosaurs inhabited the seaway.

By Kimmeridgian time, Montana sat at about 40 degrees north latitude. Deciduous trees and bushes had not yet begun to diversify, and conifers and ferns covered the land. The marine embayment had expanded farther south, leaving a broad coastal plain in Montana. Rivers flowed from uplands in the east into the embayment and deposited the stream sediments we now call the Morrison Formation. Dinosaurs, including the sauropods *Diplodocus* and *Apatosaurus*, the meat eater *Allosaurus*, and the ornithopod *Camptosaurus,* died and were preserved in this formation.

Sediments of the Morrison Formation, deposited 155 million years ago, include layers of massive sandstone alternating with thick layers of mudstone. The sandstone represents deposits from large meandering rivers, and the mudstone represents extensive river floodplains. In most of Montana, the Morrison Formation is red or green, indicating alternating episodes of oxidation and reduction and a relatively dry climate.

An extensive deposit of dinosaur bones representing a number of juvenile *Diplodocus* is preserved in the sandstone of the Morrison Formation in south-central Montana. The juveniles apparently got mired in knee-deep mud and died. The site has yielded an abundance of complete feet and lower leg bones, but the rest of the bodies have fallen apart. A site such as this gives paleontologists clues about climatic conditions, population structure, and dinosaur behavior.

The Morrison Formation is best exposed around the edges of most mountain ranges in Montana. Many outcrops of the Morrison Formation are on public land that federal agencies such as the U.S. Forest Service administer. These agencies require fossil hunters to have permits for collection.

The fossils known from the Morrison Formation in Montana

Plants
  ferns, ginkgoes, cycads, conifers
Invertebrate Animals
  freshwater mollusks
Vertebrate Animals (nondinosaurian)
  turtles
Dinosaurs
  Allosaurus, Diplodocus, Apatosaurus, ?Barosaurus, Stegosaurus, Camptosaurus

---

CRETACEOUS TIME

Early Cretaceous Time
Aptian/Albian Stages
125 to 110 million years ago

An episode of extensive erosion in Montana at the close of the Jurassic Period wiped out any evidence of geological events or dinosaur biology and evolution during that time. When deposition resumed about 30 million years later, both the geography and the animals had changed. An inland embayment extending south from the Arctic terminated in the Great Falls and Lewistown areas. The dinosaurs that lived along the rivers flowing into the embayment were very different from those of Jurassic time. *Diplodocus,* *Apatosaurus, Allosaurus,* and *Camptosaurus* had all gone extinct. Replacing them were animals such as the meat eaters *Microvenator* and *Deinonychus* and the plant eaters *Tenontosaurus* and *Sauropelta.* In early Cretaceous time, flowering plants, called angiosperms, first made their appearance.

Two formations represent early Cretaceous time in Montana: the Cloverly and the Kootenai Formations. The Cloverly Formation

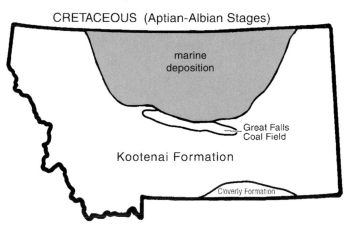

CRETACEOUS (Aptian-Albian Stages)

marine deposition

Great Falls Coal Field

Kootenai Formation

Cloverly Formation

Depositional features in Montana during the Aptian/Albian Stages of Cretaceous time

As a herd of Diplodocus travel across a mudflat, some juveniles become mired in the muck and die.
—Bill Parsons painting

© BILL PARSONS 01

yields dinosaur remains. The Kootenai Formation has rather sparse fossils and typically yields only fishes and turtles, but it has produced one poorly preserved dinosaur skeleton. The Cloverly Formation was deposited by rivers, and the sediments of the Kootenai Formation apparently accumulated in a shallow lake or possibly in part of an extension of the marine embayment.

The Cloverly Formation of south-central Montana has produced a variety of important fossils, among them a partially articulated specimen of *Tenontosaurus* that appears to have been killed and eaten by a number of *Deinonychus*. Paleontologists have found numerous *Deinonychus* teeth in the sediments directly adjacent to the *Tenontosaurus* skeleton, supporting the hypothesis that *Deinonychus* hunted in communicative groups.

The Cloverly Formation is best exposed around the Pryor Mountains of south-central Montana. Most exposures are on either Bureau of Land Management (BLM) or Crow lands, both of which require collecting permits. Like the Morrison Formation, the Kootenai Formation crops out around the edges of most mountain ranges in Montana, either on private or public lands, and permission or permits are required for collection.

The plant and animal fossils of the Kootenai and Cloverly Formations in Montana

Plants (primarily from Kootenai Formation)
   ferns, cycads, ginkgoes, conifers
Invertebrate Animals (both formations)
   freshwater mollusks
Vertebrate Animals (nondinosaurian)
(Kootenai Formation)
   fish (*Hybodus, Acrodus*), sphenodontid reptile
   (*Toxolophosaurus*)
Vertebrate Animals (nondinosaurian)
(Cloverly Formation)
   fishes (*Ceratodus*), turtles (*Naomicheleys, Glyptops*),
lizards, crocodiles, mammals (*Gobiconodon*)
Dinosaurs (primarily from Cloverly Formation)
   Deinonychus, Microvenator, megalosaur, ornithomimid,
   titanosaur, *Sauropelta, Zephyrosaurus, Tenontosaurus*

A hunting group of Deinonychus scale the neck of a titanosaur, eating the animal alive. —Bill Parsons painting

Middle Cretaceous Time
Turonian Stage
97 to 85 million years ago

The position of the continents during middle and late Cretaceous times generally resembled that of the modern world, with most of the continents separated. The Atlantic Ocean was widening as Africa separated from South America. About 97 million years ago, sea level rose and caused the Arctic Ocean embayment to join an embayment from the Atlantic that extended north from the Gulf of Mexico. The merged seaway, which geologists call the Intercontinental Cretaceous Seaway, bisected North America into two subcontinents. West America was a long, linear body of land that extended from Alaska south to southern Mexico. The Alaskan part of West America apparently connected to Siberia. The Rocky Mountains were just beginning to form. The ancient Appalachian Mountains formed the bulk of the broad land body of East America. The northern regions of East America were apparently connected from time to time to northern Europe.

Dinosaurs lived on both the eastern and western American land masses and were confined to areas within the mountain ranges or along narrow strips of coastal plains. Oceanic sediments deposited in the Intercontinental Cretaceous Seaway are called the Colorado Group, which consists of numerous marine formations. Clams, cephalopods, and reptilian plesiosaurs were the dominant life-forms of the seaway. In parts of the United States, some of the formations within the Colorado Group produce rare dinosaur remains. No dinosaur remains are known from any of the Colorado Group formations in Montana, but the Thermopolis Shale in northern Wyoming has yielded the remains of nodosaurs. In Montana, the best exposures of the Colorado Group crop out along the Marias River, in roadcuts around Shelby, and around Great Falls, Vaughn, and Ulm.

Terrestrial sediments known as the Beaverhead Conglomerate accumulated in a basin in south-

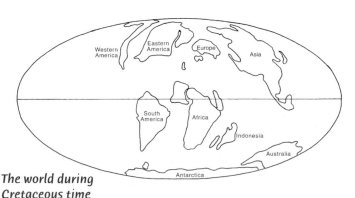

The world during Cretaceous time

western Montana during the Turonian Stage, but the rocks have not yet yielded dinosaur remains. The Beaverhead Conglomerate crops out around Dillon.

Late Cretaceous Time
Santonian Stage
85 to 80 million years ago

Eighty-five million years ago, the Intercontinental Cretaceous Seaway receded slightly, and a narrow but extensive coastal plain formed along the eastern edge of West America. In the first phase of this coastal plain's development, sediments accumulated that make up the Virgelle and Eagle Sandstones. The Virgelle Sandstone is a beach deposit that formed as the seaway receded. Nowadays you can see this beach sand preserved as the rimrocks west of Sunburst and above and around Billings. Rare, isolated dinosaur bones have been found in the Virgelle Sandstone, but they are generally worn or broken fragments. The Eagle Sandstone represents sand bodies that formed at the edge of the sea or within it. The Eagle Sandstone forms the spectacular white cliffs along the Missouri River south of Big Sandy. Like the Virgelle, the Eagle Sandstone yields rare, fragmentary dinosaur remains.

Depositional features in Montana during the early Campanian Stage of Cretaceous time

In many areas, marine deposits called the Claggett Formation overlie the Eagle Sandstone. The Claggett Formation indicates that sea level rose for a short time after the deposition of the Eagle Sandstone. The Claggett Formation has produced some isolated dinosaur bones and portions of at least one unidentified duck-bill dinosaur skeleton. Marine organisms within the formation include gastropods (snails), pelecypods, cephalopods, and various marine reptiles including plesiosaurs and mososaurs. The Claggett Formation is well exposed north of Harlem and Dodson in the Black Coulee National Wildlife Refuge, where collection permits are required, and on private ranches around Harlowton. Land-owner permission is required for collecting on ranches.

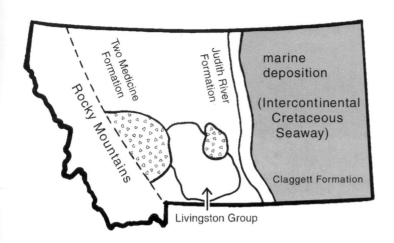

Depositional features in Montana during the middle Campanian Stage of Cretaceous time

Late Cretaceous Time Campanian Stage
80 to 74 million years ago

West of and above the Claggett Formation are two formations that represent yet another drop in sea level. The Two Medicine Formation, along the east front of the Rocky Mountains in Montana, represents the upland area of the original coastal plain. Farther east, the time-equivalent Judith River Formation represents the lowland area of the coastal plain. The Two Medicine Formation has yielded an amazing list of dinosaurs, including *Troodon, Einiosaurus, Daspletosaurus*, Montana's state fossil *Maiasaura peeblesorum*, and many more. Small streams and rivers and a few lakes deposited most of the Two Medicine Formation. Geological studies indicate that during the formation's deposition, the area was semiarid with extensive fern plains. Deciduous trees lined the waterways, and forests of conifers grew between the waterways. Herds of duck-bill and horned dinosaurs migrated across the plains. During most of the time of Two Medicine and Judith River deposition, volcanoes erupted near Helena in the area that the Elkhorn Mountains currently occupy. Volcanic ash, preserved as bentonite, is common throughout both formations, especially in the lower part of the Two Medicine Formation.

The Two Medicine Formation is 650 meters (about 2,100 feet) thick. Geologists divide it into three major subunits, informally known as the Lower, Middle, and Upper Two Medicine Formation. The lower unit was deposited as a coastal plain while the seaway was receding. Paleontologists have found primitive hadrosaurs and ceratopsians in these older sediments. The middle unit was deposited during Campanian time when sea level was at its lowest point, and the western shoreline was somewhere near the Montana–North Dakota border. During that time, *Maiasaura* and *Troodon* nested and reared their young in the uplands near the ancestral Rocky Mountains. Some *Troodon* nesting colonies existed on islands in shallow alkaline lakes. The Upper Two Medicine Formation was deposited as sea level rose with the expansion of the Intercontinental Cretaceous Seaway. *Maiasaura* was probably extinct by this time, replaced by new species of duck-bills, such as *Prosaurolophus* and *Hypacrosaurus*. Plant-eating horned dinosaurs such as *Einiosaurus* roamed the coastal plains in giant herds, and meat eaters such as *Daspletosaurus* ate the plant eaters. As sea level continued to rise, many dinosaur taxa went extinct, while others were replaced. *Achelousaurus* apparently replaced *Einiosaurus,* and around 73 million years ago, when sea level began to recede, the larger and more bizarre *Pachyrhinosaurus* apparently replaced *Achelousaurus.*

The best locations to see the Two Medicine Formation are 5 miles south of Choteau on U.S. 287 (private land) or at the Willow Creek anticline where Egg Mountain sits, 11 miles west of Choteau. The Nature Conservancy administers Egg Mountain. During the summer months, field schools operated by the Museum of the Rockies or the Teton Trail Museum in Choteau sometimes occupy the site. Daily tours of Egg Mountain allow visitors a chance to see how paleontologists study and collect dinosaur skeletons and nests. You can also see the Two Medicine Formation on the Blackfeet Reservation (permit required for collection) along U.S. 89 and around the town of Cut Bank.

In central Montana, the sedimentary deposit equivalent in age to the Two Medicine Formation is the Judith River Formation. Large meandering rivers flowing into the Intercontinental Cretaceous Seaway deposited the Judith River Formation. Much of the area was very flat, and extensive coal deposits indicate swamps and bogs probably formed within oxbow lakes. Dinosaurs included the tyrannosaur *Albertosaurus*, the duck-bill *Brachylophosaurus*, and the horned dinosaur *Avaceratops*, plus many others. The area supported relatively dense vegetation and probably looked a lot like the coast of Louisiana today—except with dinosaurs. Within the waterways lived various fishes, turtles, crocodilians, champsosaurs, and primitive mammals. Among the crocodilians was the gigantic saltwater crocodile *Deinosuchus* that exceeded 8 to 9 meters (26 to 30 feet) long and probably ate dinosaurs.

The best places to see the Judith River Formation are either along the Missouri River breaks (BLM land) or in and around the towns of Havre (private, state, and federal lands) and Harlowton (private land). The Judith River Formation represents

A Troodon colonial nesting ground
near the Rocky Mountains. These small
theropod dinosaurs brooded their eggs
using direct body contact, as birds do.
—Bill Parsons painting

two distinct depositional episodes: one as the seaway receded and another when sea level rose. You can best see the lower part of the formation, deposited during recession, north of Winifred, around the mouth of the Judith River, where Fort Claggett used to be. The upper parts of the formation crop out around Havre and Harlowton.

Fossil plants and animals known from the Two Medicine and Judith River Formations in Montana

---

UPPER TWO MEDICINE FORMATION

Plants
fossilized wood, leaf impressions of conifers and deciduous plants

Invertebrate Animals
freshwater clams, terrestrial snails

Vertebrate Animals (nondinosaurian)
fish (*Notogoneus*), amphibians (*Scapherpeton*), turtles (*Aspideretes, Basilemys*), lizards, *Champsosaurus*, crocodilians (*Leidyosuchus*), pterosaurs (*Montanazhdarcho*), mammals

Dinosaurs
Daspletosaurus, Albertosaurus, velociraptorid, *Troodon, Richardoestesia,* ornithomimid, *Orodromeus, Euoplocephalus, Edmontonia,* pachycephalosaurid, *Prosaurolophus, Hypacrosaurus,* protoceratopsian, *Achelousaurus, Einiosaurus,* birds

UPPER JUDITH RIVER FORMATION

Plants
fossilized wood, leaf impressions of conifers and deciduous plants

Invertebrate Animals
freshwater mollusks

Vertebrate Animals (nondinosaurian)
fishes (*Ceratodus, Lepisosteus, Myledaphus, Acipenser, Diphyodus*), amphibians (*Scapherpeton, Hemitrypus*),

turtles (*Adocus, Trionyx, Plastomenus, Aspideretes, Basilemys, Baena, Neurankylus*), lizards, *Champsosaurus*, crocodiles (*Leidyosuchus, Brachychampsa, Deinosuchus*), mammals (*Cimexomys, Cimolomys, Cimolodon, Meniscoessus, Mesodma, Alphadon, Pediomys, Eodelphis, Gypsonictops*)

Dinosaurs
Albertosaurus, dromaeosaurid, *Troodon*, ornithomimid, hypsilophodontid, ankylosaurid, nodosaurid, *Stegoceras, Gryposaurus*, lambeosaurine, centrosaurine, birds (*Coniornis*)

MIDDLE TWO MEDICINE FORMATION
Plants
Fossilized wood

Invertebrate Animals
freshwater clams, terrestrial snails

Vertebrate Animals (nondinosaurian)
turtles, lizards (varanoid, teiid), crocodiles, pterosaurs (Azhdarchidae), mammals (*Alphadon*)

Dinosaurs
tyrannosaurid, *Saurornitholestes, Bambiraptor, Troodon*, ornithomimid, nodosaurid, *Stegoceras, Orodromeus, Maiasaura*

LOWER JUDITH RIVER FORMATION
Plants
ferns, cycads, palms, ginkgoes, conifers, deciduous plants

Invertebrate Animals
freshwater mollusks

Vertebrate Animals (nondinosaurian)
fishes (*Lepisosteus*), turtles, *Champsosaurus*, crocodiles, mammals

Dinosaurs
tyrannosaurid, dromaeosaurid, hypsilophodontid, *Brachylophosaurus*, lambeosaurine, protoceratopsian, *Avaceratops, Ceratops*

LOWER TWO MEDICINE FORMATION

Plants
ferns, cycads, ginkgoes, conifers, deciduous plants

Invertebrate Animals
freshwater mollusks

Vertebrate Animals (nondinosaurian)
fishes (*Lepisosteus*), turtles (*Compsemys, Aspideretes*), *Champsosaurus*, crocodilians

Dinosaurs
tyrannosaurid, dromaeosaurid, *Troodon*, ornithomimid, protoceratopsian, *Gryposaurus*, ceratopsid

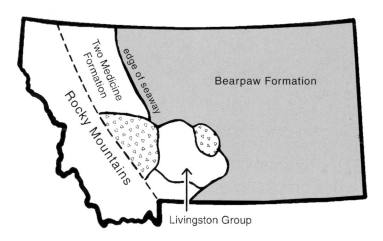

Depositional features in Montana during the Maastrichtian Stage of late Cretaceous time

Late Cretaceous Time Maastrichtian Stage
74 to 70 million years ago

The marine Bearpaw Formation overlies the Judith River Formation and much of the Two Medicine Formation. The marine shale tells us that sea level rose and narrowed the coastal plain against the Rocky Mountains. The Bearpaw Formation has produced several dinosaur skeletons that apparently washed out into the seaway as bloated carcasses. Some of these skeletons have been found articulated and preserved with skin impressions. The skeletons are typically encased in hard limestone that we must etch away using acids. Marine organisms, abundant at various levels in the formation, include clams, cephalopods, snails, crustaceans such as lobsters, various fishes, mosasaurs, and plesiosaurs.

Look for black shales of the Bearpaw Formation along the north shore of Fort Peck Reservoir from the Fred Robinson Bridge to The Pines. Extensive deposits also crop out along the Musselshell River between Melstone and Mosby. The Bearpaw Formation yields an abundance of mollusks, including ammonites, as well as marine reptiles, such as mosasaurs and plesiosaurs.

Latest Cretaceous Time Maastrichtian Stage
70 to 65 million years ago

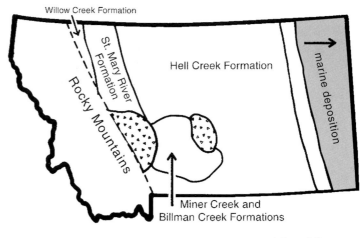

Depositional features in Montana during the Maastrichtian Stage of latest Cretaceous time

When the seaway began to recede for the last time during the Mesozoic Era, the coastal plain extended eastward until a land bridge finally reconnected East and West America. The northern and southern embayments receded— the northern one receding to the north and the southern one receding to the south—until there were no longer any shallow seas on the North American continent. As the seaways withdrew, three formations accumulated in Montana. In the west, adjacent to the Rocky Mountains, sand dunes and small streams deposited sediment under very arid conditions, and these strata we call the Willow Creek Formation. East of this desert, small streams and rivers in semiarid conditions deposited the St. Mary River Formation. A variety of dinosaurs, including *Montanaceratops*, *Hypacrosaurus*, and *Pachyrhinosaurus*, lived in this area. As the seaway moved farther east, the climate near the Rocky Mountains became drier, allowing the desert to migrate eastward. In eastern Montana, large meandering rivers flowing into the seaway deposited the Hell Creek Formation. Similar to the Judith River Formation, the Hell Creek Formation accumulated on a flat coastal plain in a warm and humid climate. The most famous dinosaurs to inhabit this area were *Tyrannosaurus*, *Triceratops*, *Edmontosaurus*, and *Pachycephalosaurus*. Some evidence suggests that *Tyrannosaurus* may have been the scavenger of the coastal plain, following giant herds of *Triceratops* or *Edmontosaurus*. Upon a carcass, the tyrannosaurs probably fought like hyenas, each trying to get as much carrion as possible. Tyrannosaurs were probably fierce, vile creatures similar to the large vultures and condors of today.

Volcanic activity around the Elkhorn Mountains continued until the end of Cretaceous time, and most of the ash is preserved in the Miners Creek, Billman Creek, and Hoppers Formations in the area surrounding the Crazy Mountains.

Two scavenger Tyrannosaurus rex eat the carcass of a Triceratops. —Bill Parsons painting

The St. Mary River and Willow Creek Formations crop out on the Blackfeet Indian Reservation (tribal and Bureau of Indian Affairs permission required) and on the Sun River Wildlife Management Area near Augusta (Montana State collection permit required). You can see the Hell Creek Formation throughout much of eastern Montana, but you can easily access it in Makoshika State Park (Montana State collection permit required) and in the Charles M. Russell Wildlife Refuge around Fort Peck Reservoir (federal permit required). In eastern Wyoming, the sediments of the same age, known as the Lance Formation, have produced similar taxa.

Sixty-five million years ago, a global disaster killed off all the nonavian dinosaurs. The geological level that records this event is called the Cretaceous/Tertiary, or K/T, boundary. You can see the boundary in Makoshika State Park in Glendive or on the road to Hell Creek State Park north of Jordan. An extensive coal layer exists at this stratigraphic level, and the level also marks a color change from the grays of the Hell Creek Formation to the tans of the overlying Paleocene sediments. One of the last dinosaurs to exist in Montana was *Tyrannosaurus rex*. Some isolated, fragmentary dinosaur bones can be found in the Paleocene sediments, which indicates that the Paleocene rivers cut down into the underlying Hell Creek Formation.

Fossil plants and animals from the St. Mary River, Willow Creek, Miners Creek, Billman Creek, Hoppers, and Hell Creek Formations in Montana

ST. MARY RIVER AND WILLOW CREEK FORMATIONS

Plants
fossilized wood

Invertebrate Animals
freshwater mollusks

Vertebrate Animals (nondinosaurian)
turtles, champsosaurs, crocodilians

Dinosaurs
Tyrannosaurid, ornithomimid, ?*Hypacrosaurus*, hadrosaurine, *Montanaceratops*

MINERS CREEK, BILLMAN CREEK, AND HOPPERS FORMATIONS

Plants
fossil wood

Invertebrate Animals
freshwater mollusks

Vertebrate Animals (nondinosaurian)
turtles

Dinosaurs
tyrannosaurid, *Edmontosaurus*

HELL CREEK FORMATION

Plants
fossil wood, cycads, palms, conifers, deciduous plants

Invertebrate Animals
freshwater mollusks

Vertebrate Animals (nondinosaurian)
fishes (*Lonchidion, Ischyrhiza, Myledaphus, Acipenser, Protoscaphirhynchus, Paleopsephurus, Amia, Melvius, Belonostomus, Palaeolabrus, Phyllodus, Parabula, Lepisosteus, Coriops, Platacodon*), amphibians (*Scapherpeton, Lisserpeton, Prodesmodon, Opisthotriton, Palaeobatrachus, Proamphiuma, Habrosaurus, Scotiophryne, Eopelobates*), lizards (*Chamops, Leptochamops, Haptosphenus, Peneteius, Contogenys, Pancelosaurus, Exostinus, Colpodontosaurus, Parasaniwa, Paraderma, Palaeosaniwa*), *Champsosaurus*, crocodilians (*Leidyosuchus, Brachychampsa*), mammals (*Mesodma, Cimexomys, Cimolodon, Catopsalis, Stygimys, Cimolomys, Meniscoessus, Essonodon, Alphadon, Glasbius, Pediomys, Didelphodon, Gypsonictops, Batodon, Cimolestes, Procerberus, Protungulatum*)

Dinosaurs
Tyrannosaurus, dromaeosaurid, *Troodon, Paronychodon, Richardoestesia*, caenagnathid, *Ornithomimus, Thescelosaurus, Buganosaura, Stegoceras, Pachycephalosaurus, Stygimoloch, Leptoceratops, Triceratops, Torosaurus, Ankylosaurus, Edmontosaurus, Anatotitan, Avisaurus*, and birds

Dinosaurs and Other
Mesozoic Fossils of Montana

THIS CHAPTER'S AIM is to help amateur and professional paleontologists identify some of the more common bones they might find in the Mesozoic formations of Montana. It describes some of the interesting information we think we've discovered about many of the dinosaur fossils found in and near Montana. I have organized this chapter by geological formation and time. It includes all the dinosaur taxa that have been identified from each formation with descriptions of some common elements. Bone drawings from some of my notebooks illustrate the more common skeletal remains and teeth of dinosaurs you might find under the Big Sky. Most rock formations that contain dinosaur remains hold many other fossils as well, so I have included illustrations of some of the more common nondinosaurian remains that commonly accompany dinosaur fossils in microsites. For some of these bones and teeth, I've added scales to give you some idea of their size. Keep in mind that dinosaurs of different growth stages have different sizes of bones. For example, the humerus of a duck-bill dinosaur looks identical whether it is from a hatchling or an adult. The only difference is its size. Scale is not a very reliable tool in identifying a dinosaur bone or tooth, so the scales I've provided may be of only limited help.

The names of the dinosaurs that are in italics are the Latin or Greek genus and species names. Those scientific names not in italics are family or subfamily names. A question mark preceding a scientific name indicates uncertainty in the identification. The words in quotations are the English translations of the scientific names. In parentheses is the pronunciation for some of the tongue-twisting scientific names.

The job of naming a new fossil specimen, such as a new dinosaur, falls to the person who first does scientific research about the specimen, or the person who decides the specimen is worthy of being named a new species. In 1978, my late friend Bob Makela and I excavated an adult duck-bill dinosaur skull that Laurie Trexler had located in the Two Medicine Formation of western Montana. When we cleaned the rock off the skull, we could tell it was not only a new species but also a new genus of duck-bill. It was simply different from anything else that anyone had ever described—so we described and named it. Because baby duck-bill dinosaurs had been found nearby and the skulls of the babies had features similar to those of the adult skull, we hypothesized that both the babies and adult were the same species. The babies were found in the confines of a nest and looked like they had been there for some time after they hatched. Based on these observations, Bob and I hypothesized that these dinosaurs cared for their young. For help in naming this dinosaur, I went to my boss at the time, Don Baird, who has a tremendous understanding of scientific nomenclature and all the rules of naming new species. Don thought for a day and came up with the name *Maiasaura,* meaning "good mother reptile."

There is no particular rule that someone naming a new species must follow other than that they must select a name that is either Greek or Latin. However, it is best if the name also is descriptive. In the case of *Maiasaura,* the name refers to the parental care hypothesis. The whole name of the dinosaur is *Maiasaura peeblesorum.* The species name, *peeblesorum,* is the Latin way of honoring the Peebles families, owners of the land where we collected the specimens.

Scientific names are supposed to be scientific, but sometimes people create names that are also silly. For example, *Cuttysarkosaurus* is a lizard that one of my colleagues named—perhaps after he had been out in the field too long. Some dinosaur names get long when people try to include too much description—for instance, the dinosaur from Asia called *Micropachycephalosaurus.* I think it might be the longest dinosaur genus name. And some dinosaurs, particularly some from China, have names that are hard to pronounce. Try wrapping your mouth around *Zizhongosaurus* and *Xuanhanosaurus.* I have no idea of the derivation of these names, nor do I know how to pronounce them. There are hundreds of dinosaur names that I don't know how to pronounce, including

Tuojiangosaurus, Unquillosaurus, and *Wuerhosaurus.* The two dinosaur genera I've been responsible for, *Maiasaura* and *Orodromeus,* seem so simple, but someone whose native tongue is a language other than English might find those names as hard to pronounce as *Kakuru kujani* is for me.

I have listed the museum number of the type specimen for some of the dinosaurs. When scientists first name extinct animals and plants, the original specimen that is described is designated the holotype, or type specimen. Researchers consult the type specimen and its description when attempting to identify another similar fossil. It is very important that the type specimen remain in a museum collection with public access because researchers are constantly studying these specimens. Important specimens, such as type specimens, in private hands risk being transferred or lost. For example, the type specimen of *Bambiraptor* was collected commercially and sold to a private individual. The specimen is now on display at a museum in Florida, but it is unclear whether it will remain there. In cases where the type specimen resides at a museum or repository outside Montana, I have included a reference specimen from the Museum of the Rockies collection. You will find data about these reference specimens on the Museum of the Rockies Paleontology Department web site at www.museum.montana.edu. The type and referred specimens are from various museums and therefore carry different letter and number designations. The letters signify the museum that houses the type specimen.

The abbreviations used in type specimen numbers and the museums they represent

AMNH = American Museum of Natural History, New York, New York

ANSP = Academy of Natural Sciences, Philadelphia, Pennsylvania

CM = Carnegie Museum of Natural History, Pittsburgh, Pennsylvania

CMNH = Cleveland Museum of Natural History, Cleveland, Ohio

FIP = Florida Institute of Paleontology,
Dania Beach, Florida

MCZ = Museum of Comparative Zoology,
Harvard University, Cambridge, Massachusetts

MOR = Museum of the Rockies,
Montana State University, Bozeman, Montana

NMC = National Museum of Canada,
Ottawa, Ontario, Canada

OTM = Old Trail Museum, Choteau, Montana

ROM = Royal Ontario Museum,
Toronto, Ontario, Canada

SDSM = South Dakota School of Mines,
Rapid City, South Dakota

UCMP = University of California Museum of
Paleontology, Berkeley, California

UM = University of Montana, Missoula, Montana

USNM = United States Museum of Natural History
(Smithsonian Institution), Washington, D.C.

YPM = Yale Peabody Museum, Yale University,
New Haven, Connecticut

YPM-PU = Yale Peabody Museum (Princeton University
Collection), Yale University, New Haven, Connecticut

The following abbreviations in the captions to photographs denote the management of the land where the specimen was found

ACE = Army Corps of Engineers
BfN = Blackfeet Nation
BLM = Bureau of Land Management
MT = Montana State Lands
NC = The Nature Conservancy
Pvt = Private land
USFW = U.S. Fish and Wildlife Service

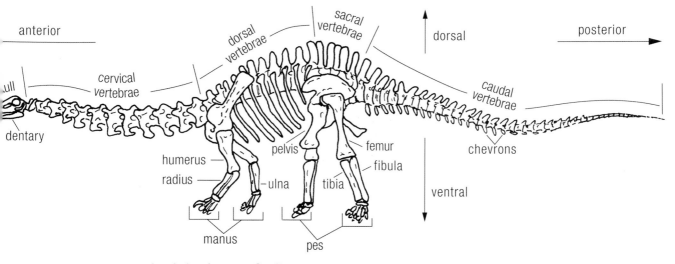

The skeletal parts of a dinosaur, as seen in a sauropod skeleton, and some anatomical directions we use to get our orientation

dentary = lower jaw
cervicals or cervical vertebrae = neck vertebrae
dorsals or dorsal vertebrae = vertebrae of the body
sacrals or sacral vertebrae = vertebrae of the pelvic region
caudal vertebrae = vertebrae of the tail
chevrons = spines that protrude below the caudal vertebrae
humerus = upper arm, or front leg bone
ulna = lower arm, or front leg bone that forms elbow
radius = lower arm or front leg bone
manus = hand or front foot
femur = upper hind leg bone
tibia = one of two lower hind leg bones
fibula = one of two lower hind leg bones
pes = hind foot. Toe bones of the manus and pes are called phalanges, and the last toe bones are called unguals.

JURASSIC DINOSAURS OF MONTANA

The terrestrial Jurassic rocks of Montana are the Morrison Formation. This formation differs somewhat from the rocks of the same name in Wyoming, Colorado, and Utah. In Montana, the formation is considerably thinner and has not been the focus of many paleontological studies.

In Montana, marine rocks of the Middle Jurassic Swift Group lie beneath the Morrison Formation, and the Cretaceous Cloverly or Kootenai Formations lie above it. The Jurassic-Cretaceous contact is an erosional unconformity, which records a long period of erosion during which no deposition took place. Most of the exposures of the Morrison Formation are around the edges of mountain ranges. As a result, the rocks typically are steeply tilted, making excavation difficult. Most of our knowledge of Montana's Jurassic paleontology comes from studies of the underlying marine Swift Group. Rocks of the Swift Group contain a variety of invertebrate fossils, including belemnites, crinoids, and various brachiopods. These strata have also yielded fish skeletons.

LATE JURASSIC
Morrison Formation
Kimmeridgian Stage
153 million years ago

Upper Jurassic sediments are not common in Montana, but in some of their rare exposures, collectors have found dinosaur remains, often in abundance. Sauropod bones are the most common remains in the Morrison Formation, and because most sauropods grew to enormous sizes, their bones are also large. Most of the bones from the Morrison Formation are black if found partially or completely covered by sediments. Specimens that have been weathering out of the rock for some time may be bluish gray, gunmetal blue, or even bleached to tan or white.

Most of the deposits of bones in this region appear to represent skeletal accumulations deposited in river systems over time. Two deposits that have been located and partially collected, however, appear to be accumulations that may represent mass death sites. Interestingly, these two sites produce the remains of juvenile diplodocid sauropods, suggesting that these young dinosaurs may have lived in social groups.

Collectors have not yet discovered the bones of many of the smaller ornithischian species in Montana, but this is likely because of limited exploration. There is still a great deal of work to be done in the Morrison Formation of Montana.

Time position of the Morrison Formation

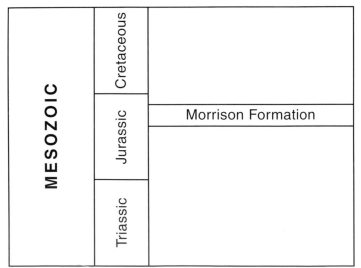

Ornithischian Dinosaurs

Camptosaurus sp. (species unidentified)

Camptosaurus (CAMP-toe-sore-us) = "bent reptile"

Referred Specimen: MOR 695 (partial skeleton)

Camptosaurus was a medium, bipedal ornithopod (Ornithischian) dinosaur that grew to about 5 meters (17 feet) long. *Camptosaurus* had a row of grinding teeth for chewing plants. All ornithischian dinosaurs are easily identified because they have a predentary bone, or "beak," that connects to their lower jaws. *Camptosaurus* is closely related to *Iguanodon* and shares an ancestry with the hadrosaurs, or duck-bill dinosaurs.

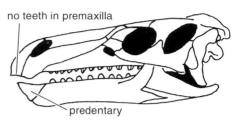

no teeth in premaxilla

predentary

Camptosaurus skull (left lateral view)

Camptosaurus teeth have enamel on one side, and narrow ridges run down the height of the enamel crown. The crown of an average adult dentary tooth—that is, a tooth in the lower jaw—is about 1.5 centimeters (0.6 inch) long. Collectors might confuse *Camptosaurus* teeth with those of *Tenontosaurus* from the Lower Cretaceous Cloverly Formation. Because the Morrison and Cloverly Formations commonly crop out in the same area, fossil hunters need to carefully determine the stratigraphic level of their discovery.

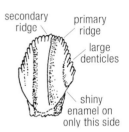

secondary ridge

primary ridge

large denticles

shiny enamel on only this side

Camptosaurus dentary tooth

The hind feet of *Camptosaurus* have vestigial first digits. The phalanges of the other digits are flattened and widened laterally, similar to *Iguanodon* and the hadrosaurs. The claws of the foot, called the pes unguals, are more pointed than wide, making the feet of *Camptosaurus* appear clawed rather than hooflike.

The caudal vertebral centra have striations along their lateral surfaces, similar to although not as pronounced as *Thescelosaurus* from the Hell Creek Formation.

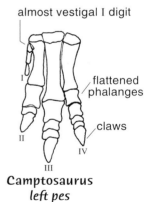

almost vestigal I digit

flattened phalanges

claws

I

II

III

IV

Camptosaurus left pes

Camptosaurus remains are also found in Wyoming, Colorado, Utah, and South Dakota. Their remains are very rare in Montana, probably because so little exploration has been conducted in the Morrison Formation. A single fragmentary skeleton is known from Montana, and it is apparently a juvenile.

Camptosaurus caudal vertebral centrum

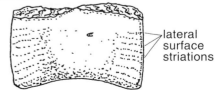

lateral surface striations

Stegosaurus

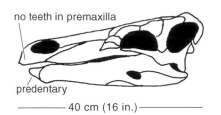

no teeth in premaxilla

predentary

——— 40 cm (16 in.) ———

Stegosaurus skull

1.5 cm
(0.6 in.)

round root

Stegosaurus
tooth (side view)

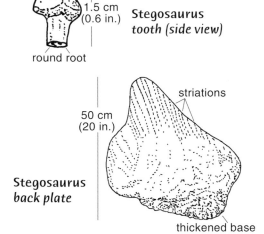

striations

50 cm
(20 in.)

Stegosaurus
back plate

thickened base

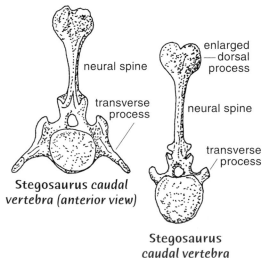

neural spine

transverse
process

Stegosaurus caudal
vertebra (anterior view)

enlarged
dorsal
process

neural spine

transverse
process

Stegosaurus
caudal vertebra
(posterior view)

Stegosaurus sp. (species unidentified)
Stegosaurus (STEG-o-sore-us) = "roof reptile"

Stegosaurus was an armored ornithischian dinosaur with large plates on its back that it used for display and possibly for defense. Some researchers have hypothesized that stegosaurs used their plates for heat regulation, but there is little evidence to support the theory. Like many dinosaurs, *Stegosaurus* had a very small skull for its body size and a particularly small brain—no larger than a couple of walnuts. Apparently, brains were not very important during Jurassic time.

Stegosaurus remains have been reported from Montana but are rare. Stegosaur teeth closely resemble those of most other primitive ornithischians in their small size and leaflike shape. The animals apparently used them for nipping but not for chewing—the teeth are seldom very worn as would be expected if the animals had used them to grind plant material. Stegosaur teeth are about the size of the end of your little finger.

Portions of stegosaur armor plates are the most commonly found remains of these animals. Whole plates can measure as much as 70 or 80 centimeters (27 to 31 inches) long and 60 to 70 centimeters (23 to 27 inches) wide but are generally no more than about 2 or 3 centimeters (about 1 inch) thick. You can identify pieces of plates by their thickness and a flat surface commonly covered with shallow grooves. These grooves may have carried blood vessels to a sheath covering. Tail spikes also have shallow grooves. Most stegosaur vertebrae are easy to identify because of the enlarged dorsal process on the ends of the neural spines. These enlargements apparently helped support the overlying plates.

Most *Stegosaurus* remains have been found in the Morrison Formation of Wyoming, Colorado, and Utah. There are no *Stegosaurus* remains in the Museum of the Rockies repository, although a partial skeleton, found south of Great Falls, does exist in a private collection.

Saurischian Dinosaurs

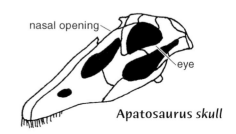

Apatosaurus

Apatosaurus sp. (species unidentified)
Apatosaurus (ah-PAT-o-sore-us) = "deceptive reptile"
Referred Specimen: MOR 592
(partial skeleton)

Apatosaurus is the dinosaur that we used to call *Brontosaurus*. Its name has changed because of a scientific rule that says that the first name given to an animal has priority over any subsequent names. Charles O. Marsh named *Apatosaurus* in 1877 on the basis of a sacrum and some vertebrae. In 1879, he named *Brontosaurus* on the basis of the major portion of a skeleton. Marsh believed that *Brontosaurus* and *Apatosaurus* were two different animals. The *Brontosaurus* skeleton became very famous, and almost everyone knew about it. But studies by Dave Berman and Jack McIntosh in 1978 revealed that *Brontosaurus* and *Apatosaurus* were the same species. So the name *Apatosaurus* replaced the name *Brontosaurus*.

An average adult *Apatosaurus* was about 25 meters (80 feet) long and weighed as much as 35 tons. Its remains are most common in the Morrison Formation of Colorado, Utah, and Wyoming. Only disarticulated remains, primarily from bonebeds, have been found in Montana.

You can identify the different species of sauropod dinosaurs (brontosaurs or long-necks) only if you have the right parts of the skeleton. For most dinosaurs the skulls are the easiest parts to identify. But the skulls of *Apatosaurus* and *Diplodocus* look very similar, so the vertebrae are generally used to identify them. The skulls of both *Apatosaurus* and *Diplodocus* are elongated with the nasal, or nose, opening on top of their heads. The teeth of *Apatosaurus* look like those of *Diplodocus* in that they are pencil- or peg-shaped. *Apatosaurus* had teeth only in the front of its mouth. The teeth are about the size of your thumb.

A good way to identify *Apatosaurus* is with one of its anterior caudal vertebrae. These vertebrae can be very large. The neural spines of these bones have a massive, square, enlarged knob at

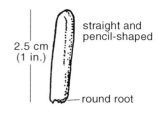

Apatosaurus *tooth*

Apatosaurus skull

nasal opening

eye

2.5 cm
(1 in.)

straight and pencil-shaped

round root

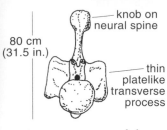

knob on neural spine

80 cm (31.5 in.)

thin platelike transverse process

Apatosaurus caudal vertebra (anterior view)

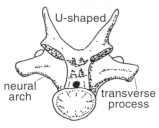

U-shaped

neural arch

transverse process

Apatosaurus dorsal vertebra (posterior view)

the upper end of the spine. The anterior dorsal vertebrae are also identifiable because the split neural arch is very wide open compared with those of most other sauropod dorsals.

If you are collecting in sediments of the Morrison Formation and you come across bones as big as a person, chances are pretty good that they belong to a sauropod.

Diplodocus

Diplodocus sp. (species unidentified)

Diplodocus (dih-PLO-doe-kus) = "double beam"

Referred Specimens: MOR 790 (bonebed)

Diplodocus was a very slender and gracile sauropod that reached as much as 27 meters (88 feet) long. Researchers think it weighed about 11 tons. Traditional reconstructions of *Diplodocus* and other sauropods have them eating the leaves of tall conifer trees. In recent studies, paleontologists suggest that the animal could not reach this elevation and instead that the dinosaur used its long neck to browse over a large area without having to move its giant body.

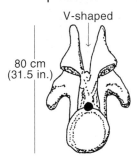

nasal opening (nose)

40 cm (16 in.)

Diplodocus skull

You can identify *Diplodocus*, like *Apatosaurus*, on the basis of its dorsal and caudal vertebrae. They have a split, or double, neural spine or beam—hence the Latin name. These vertebrae can be very large, often measuring 1 meter (3 feet) tall. The neural spines of the anterior caudal vertebrae do not have an enlarged knob. The split neural arches of the anterior dorsal vertebrae are narrow and V-shaped.

Diplodocus teeth are shaped like blunt pencils.

Remains of *Diplodocus* and *Apatosaurus* are often found in the same deposits, suggesting that they lived together. Most of the *Diplodocus* remains from Montana are represented by juveniles found in bonebeds. One of the bonebeds suggests that many of the juveniles may have been stuck knee-deep in mud—excavations revealed the lower extremities of their legs preserved more often than other parts. Articulated feet are common, but articulated vertebrae are rare. Many of the feet and lower leg bones are upright in the sediments, suggesting that the animals got stuck rather than being washed in during a flood.

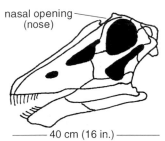

V-shaped

80 cm (31.5 in.)

Diplodocus anterior dorsal vertebra (posterior view)

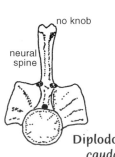

no knob

neural spine

Diplodocus anterior caudal vertebra

Allosaurus fragilis MARSH, 1877

Allosaurus (AL-o-sore-us) = "different reptile";
fragilis (FRA-jill-is) = "brittle"
Type Specimen: YPM 1930
Referred Specimen: MOR 693
(articulated skeleton)

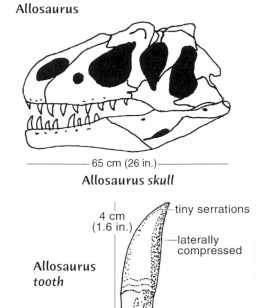

Allosaurus

Allosaurus *skull*

————— 65 cm (26 in.) —————

Allosaurus was a meat-eating dinosaur that reached 8 to 9 meters (26 to 29 feet) long and may have weighed 1 to 2 tons. *Allosaurus* was one of the largest meat eaters living during late Jurassic time in North America and was probably the primary predator on the sauropods.

Allosaurus had short hornlike structures over each eye, and a very narrow skull with laterally compressed teeth in its jaws. *Allosaurus* had relatively short arms with enlarged hand claws that are strongly recurved. The claws can be as large

4 cm (1.6 in.)

tiny serrations

laterally compressed

Allosaurus tooth

Allosaurus skull from the Morrison Formation of northern Wyoming (BLM).
Skull is 75 centimeters (29 inches) long.

blood
groove

*Allosaurus ungual
phalange from manus*

*Allosaurus
pes ungual*

spindle-
shaped

*dorsal vertebra
(lateral view)*

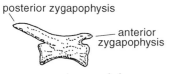

posterior zygapophysis

anterior
zygapophysis

*posterior caudal
vertebra*

as a person's hand. The claws were probably useful for hanging on to live prey or ripping flesh during feeding. *Allosaurus* has a three-fingered manus, or hand.

Allosaurus remains, including their teeth, are commonly found in river deposits associated with the bones of sauropods and other plant eaters. The teeth, generally black and shiny, measure 4 or 5 centimeters (1.5 to 2 inches) long, with very small serrations (about 27 per centimeter or about 10 per inch). Like most other carnivorous dinosaurs, the dorsal vertebral centra are extremely spindle-shaped, and the caudal vertebrae are elongated with long zygapophyses. The claws from the manus are large and strongly curved.

Allosaurus remains are uncommon in Montana, consisting mainly of isolated teeth and bones, commonly in sauropod bonebeds. The Cleveland-Lloyd Quarry in Utah has produced numerous skeletons of varying sizes of *Allosaurus,* suggesting that these dinosaurs may have hunted in groups or at least that they were social animals. A nearly complete skeleton of *Allosaurus* found in northern Wyoming has pathologies on many of its bones. These indicate that the animal had severe infections, some that may in the end have killed the creature.

Nondinosaurian Vertebrate Fossils from the Morrison Formation of Montana

Nondinosaurian remains are rare from the Morrison Formation, probably because there have been so few dinosaur excavations. One site yielded a few fragmentary turtle remains together with an associated sauropod skeleton.

CRETACEOUS DINOSAURS OF MONTANA

Most of the Mesozoic rocks of Montana accumulated during Cretaceous time between 112 million years ago and 65 million years ago. During this period, sea level changed rhythmically, rising and falling repeatedly. As a result, marine formations separate the Cretaceous terrestrial sediments that yield dinosaur remains from one another. Most dinosaur remains are found in terrestrial rocks, but a few have been found in the marine strata. The dinosaur skeletons that Earl Douglass collected in 1900 came from marine strata of the Bearpaw Formation, and Barnum Brown reported a nodosaur specimen from the Thermopolis Shale. However, most Cretaceous dinosaur remains come from nonmarine sediments, mainly on the east side of the Continental Divide.

EARLY CRETACEOUS
Cloverly Formation
Aptian/Albian Stages
112 million years ago

The Cloverly Formation consists mainly of red, purple, and gray mudstones with a few brown sandstones. John Ostrom identified three stratigraphic units within the formation, each with slightly different geological characteristics. Most of the articulated and associated dinosaur fossils have been discovered in the upper, colored layers, but some specimens, including dinosaur egg and baby remains, have come from the lower, gray bentonitic beds and caliche deposits. Preservation in the lower sediments is generally poor, and skeletal elements are typically distorted.

	Cretaceous	Cloverly Formation
MESOZOIC	Jurassic	
	Triassic	

Time position of the Cloverly Formation

The Jurassic Morrison Formation lies below the Cloverly Formation, and black shales of the marine Thermopolis Formation lie above. Both contacts are erosional.

Ornithischian Dinosaurs

Tenontosaurus tilletti OSTROM, 1970
Tenontosaurus (ten-ON-toe-sore-us) = "sinew reptile"; *tilletti* (till-et-eye) = in honor of the Lloyd Tillett family of south-central Montana and Lovell, Wyoming
Type Specimen: AMNH 3040
Discovered by: Barnum Brown
Referred Specimen: MOR 682 (articulated skeleton)

very long tail

Tenontosaurus

Tenontosaurus was a plant-eating dinosaur closely related to *Iguanodon*. It was probably one of the favorite foods of *Deinonychus*, because several *Tenontosaurus* carcasses have been found with *Deinonychus* teeth around them. *Tenontosaurus* was bipedal but could put its hands down to help when in a slow walk. *Tenontosaurus*

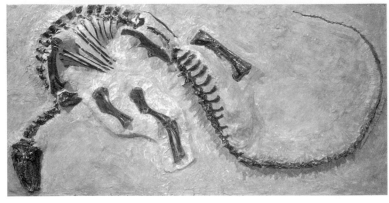

Skeleton of Tenontosaurus *from the Cloverly Formation of south-central Montana (BLM). Skull is 33 centimeters (13 inches) long.*

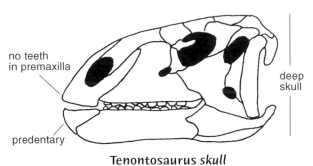

no teeth in premaxilla

deep skull

predentary

Tenontosaurus *skull*

Skull of Tenontosaurus *from the Cloverly Formation of south-central Montana (BLM). Skull is 33 centimeters (13 inches) long.*

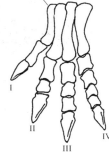

large digit I compared with *Camptosaurus*

I

II

III

IV

Tenontosaurus
left pes

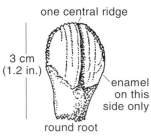

one central ridge

3 cm
(1.2 in.)

enamel on this side only

round root

Tenontosaurus *tooth*

reached nearly 8 meters (26 feet) long and may have weighed close to 1 ton, although an average *Tenontosaurus* was only about 5 meters (16 feet) long. *Tenontosaurus*, like other primitive ornithischians, had a very long tail for the size of the animal. As a result, caudal vertebrae are relatively common in the Cloverly Formation.

Tenontosaurus is known from several skeletons, including parts of tiny babies. Most of the skeletons collectors have found were articulated, or intact, suggesting that sediments had swiftly covered their bodies. Some juvenile skeletons have been found together, indicating that they may have lived in sibling groups.

Skeletal elements of *Tenontosaurus* are probably the most common dinosaur remains found in the Cloverly Formation of Montana and are commonly articulated. *Tenontosaurus* teeth are recognizable. They have enamel only on one side and a central ridge that runs down the length of the enameled surface, with several small ridges alongside the central ridge. An adult *Tenontosaurus* tooth is about the size of a grown person's thumb.

Tenontosaurus vertebral centra are round and about the size of a person's fist. Collectors commonly find articulated *Tenontosaurus* tails because massive bundles of ossified tendons apparently held them together long after the animals died. The femur of an average adult *Tenontosaurus* is about 35 centimeters (14 inches) long, and 6 centimeters (about 2 inches) in diameter. The humerus has a very large deltopectoral crest and resembles the humerus of a hadrosaur, or duck-bill dinosaur.

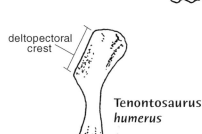

Tenontosaurus anterior caudal vertebra — tall spines

Tenontosaurus posterior caudal vertebra

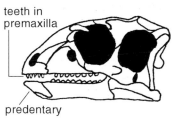

deltopectoral crest

Tenontosaurus humerus

Zephyrosaurus schaffi SUES, 1980

Zephyrosaurus (ZEF-i-roe-sore-us) = "west wind reptile"; schaffi (shaf-i) = to honor paleontologist Chuck Schaff
Type Specimen: MCZ 4392
Discovered by: Chuck Schaff
Referred Specimen: MOR 759
(associated skeleton)

Zephyrosaurus was a small, bipedal, plant-eating hypsilophodontid very closely related to the late Cretaceous hypsilophodontid *Orodromeus*, known from Egg Mountain. *Zephyrosaurus* remains are very rare—only two relatively good specimens exist and neither is complete. *Zephyrosaurus* was only about 1 meter (3 feet) long and probably weighed no more than about 11.2 kilograms (25 pounds).

Vertebrae are about the size of the end of a person's thumb. The anterior caudal vertebrae have long, bladelike transverse processes that connect to the vertebrae at the junction between the neural arch and centrum. The anterior and posterior surfaces of the bottom (ventral) and sides (lateral) of the dorsal centra have roughened striations

Zephyrosaurus

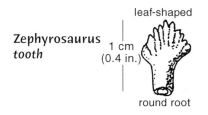

teeth in premaxilla

predentary

Zephyrosaurus skull

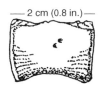

leaf-shaped

1 cm (0.4 in.)

round root

Zephyrosaurus tooth

— 2 cm (0.8 in.) —

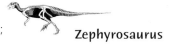

Zephyrosaurus caudal vertebra centrum

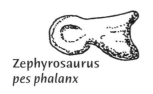

Zephyrosaurus
pes phalanx

Sauropelta

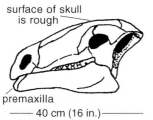

keel

12 cm
(4.7 in.)

Sauropelta
armor plate

surface of skull
is rough

premaxilla
—— 40 cm (16 in.) ——

Sauropelta skull

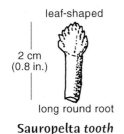

leaf-shaped

2 cm
(0.8 in.)

long round root

Sauropelta tooth

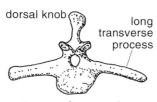

dorsal knob

long
transverse
process

Sauropelta anterior
caudal vertebra

similar to those in the Upper Cretaceous ornithopod *Thescelosaurus* of the Hell Creek Formation.

An adult *Zephyrosaurus* tibia is about 21 centimeters (8 inches) long. The third metatarsal is 9.5 centimeters (3.7 inches) long. *Zephyrosaurus* is known only from Montana.

Sauropelta edwardsi OSTROM, 1970

Sauropelta (sore-o-PELT-a) = "small shield reptile";
edwardsi = to honor Nell and Tom Edwards of Bridger, Montana
Type Specimen: AMNH 3032
Referred Specimen: MOR 345 (partial skeleton)

Sauropelta was a quadrupedal, plant-eating, armored dinosaur measuring about 5.5 meters (18 feet) long and weighing about 3 tons. It had a massive, gnarly skull, and bony armor plates covered its body. *Sauropelta* armor is relatively common in the Cloverly Formation. The armor plates are usually about the size of a person's open hand and have a rough and pitted surface. A thick horny sheath probably covered the armor plates. The dinosaurs most likely used their armor to protect themselves from predators.

Like other armored dinosaurs, *Sauropelta* had massive ribs that were fused or united to their vertebrae. Armored dinosaur caudal vertebrae are distinct because they have very large processes for the chevrons, and the chevrons are fused to the vertebral centrum. The pelvis of *Sauropelta* was fused together as a rigid structure to support the heavy body.

The teeth of *Sauropelta* are leaf-shaped with round roots, closely resembling those of other armored dinosaurs, including *Stegosaurus*. Because no other kinds of armored dinosaurs are found in the Cloverly Formation, the leaf-shaped teeth of *Sauropelta* are easy to identify.

Sauropelta is known from numerous articulated skeletons from Montana and Wyoming. The skeletons are often found preserved with carbonate caliche deposits surrounding the bones. These carbonate deposits are extremely hard and usually can only be removed safely by using an acid solution such as glacial acetic acid.

Saurischian Dinosaurs

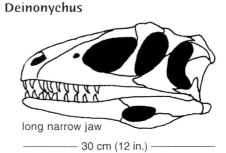

Deinonychus

Deinonychus antirrhopus OSTROM, 1969

Deinonychus (die-NON-uh-kus) = "terrible claw";
antirrhopus (an-tih-ROE-pus) = refers to the
hook-shaped *(rho)* claw of the foot *(pus)* directed
in the opposite *(anti)* way

Type Specimen: YPM-5205
Discovered by: John Ostrom, 1964
Referred Specimen: MOR 747 (bonebed)

long narrow jaw

———— 30 cm (12 in.) ————

Deinonychus skull

Deinonychus is one of the most famous dinosaurs—in the movie *Jurassic Park,* the raptors were deinonychids. *Deinonychus* was probably one of the fiercest animals to have ever lived. Like *Velociraptor, Deinonychus* killed its prey by seizing it with razor-sharp hand claws and then ripping the animal open with the slashing claws on its hind feet. Evidence from sites in Montana suggest that *Deinonychus* hunted in packs of three to five individuals. Adult *Deinonychus* skeletons are about 2.5 meters (8 feet) long. At about 79 kilograms (175 pounds), *Deinonychus* was about the size of an adult person.

Mounted Deinonychus *skeleton from the Cloverly Formation of south-central Montana (MT). Skull is 35 centimeters (about 14 inches) long.*

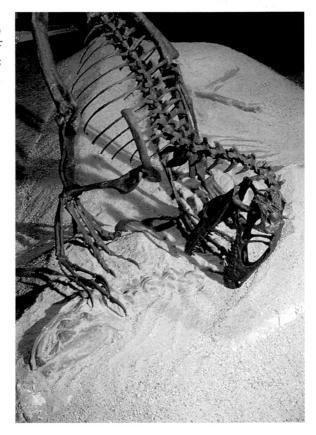

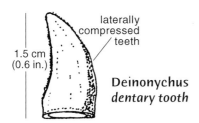

1.5 cm (0.6 in.)

laterally compressed teeth

Deinonychus *dentary tooth*

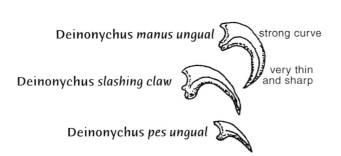

Deinonychus *manus ungual* strong curve

Deinonychus *slashing claw* very thin and sharp

Deinonychus *pes ungual*

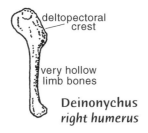

deltopectoral crest

very hollow limb bones

Deinonychus right humerus

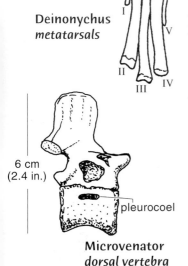

Deinonychus metatarsals

I

V

II

III

IV

6 cm (2.4 in.)

pleurocoel

Microvenator dorsal vertebra

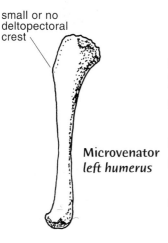

small or no deltopectoral crest

Microvenator left humerus

Deinonychus remains are very rare, but you can't mistake their teeth and claws, as long as you are collecting from the Cloverly Formation. The teeth are about 1 to 2 centimeters (about 0.5 inch) long and have serrated edges. The hand claws and sickle claw on each hind foot are strongly recurved and flattened from side to side, indicating that in life they were extremely sharp. In life, they would have been covered by an even larger horny claw-sheath. *Deinonychus* had long arms and long fingers for grasping prey. The leg bones of *Deinonychus* are hollow, like those of a bird. The dorsal vertebrae have holes, called pleurocoels, on either side of the centrum, like those of most other carnivorous dinosaurs. Long bundles of rod-shaped bones surrounded the tail vertebrae and kept the tail relatively rigid.

Remains of *Deinonychus* are rare because carnivorous animals are less common than herbivores and because the bones of carnivorous dinosaurs were hollow, lightly built, and therefore subject to breakage. The fragmentary remains of fewer than ten *Deinonychus* individuals are known. Most of these are from Montana.

Microvenator celer OSTROM, 1970
Microvenator (MY-crow-ven-AY-tore) = "small hunter";
celer (SELL-er) = "swift"
Type Specimen: AMNH 3041

Microvenator was a small, lightly built, meat-eating coelurosaurian dinosaur probably no more than 1 meter (about 3 feet) long. *Microvenator* remains are extremely rare, and only one partial skeleton is known.

The femur of *Microvenator* is about 13.5 cm (5.3 inches) long, and the tibia is about 16.5 cm (6.4 inches) long. The femur is strongly curved, and the tibia is very straight.

Microvenator is now thought to have been an oviraptorid, very closely related to birds. Oviraptorids lack teeth.

Ornithomimus sp. (species unidentified)
Ornithomimus (or-nith-o-MIME-us) = "bird imitator"

All ornithomimids share a similar shape and look much like the gallimimuses running in the herd in *Jurassic Park*. John Ostrom identified the "ostrich dinosaur" bones from the Cloverly

Ornithomimus

Formation as being *Ornithomimus*. However, once someone finds more complete material, we likely will discover that this is a new genus of ornithomimid, and not *Ornithomimus*.

The most common fossils of ornithomimids are their elongated toe bones, called phalanges. You can identify the toe bones by their flattened bottom, or ventral, surfaces, and by the their uncurved claws.

Ornithomimid bones are very rare in the Cloverly Formation.

Titanosauridae (genus and species unidentified)
Titanosaurid (ty-TAN-o-sore-id)
Referred Specimen: MOR 334 (associated partial skeleton)

Titanosaurs are a group of sauropod dinosaurs that were relatively common during Cretaceous time. Titanosaur bones, although rare in the Cloverly Formation, are easy to identify because of their size. A titanosaur vertebra can be as large as a basketball, and the femur can be more than 1.5 meters (5 feet) long. The Cloverly titanosaurid was probably about 13 meters (43 feet) long, and may have weighed around 6 tons. Titanosaur teeth are blunt and about as big around as a pencil.

Some paleontologists have identified extremely shiny, polished rocks found in the Cloverly Formation as the gastroliths, or crop stones, of titanosaurs, but the identification of the rocks as crop stones is still controversial. Other sauropods may have had these stomach stones as well.

Common Nondinosaurian Fossils from the Cloverly Formation

Occasionally collectors find the teeth of lungfishes, *Ceratodus frazieri,* in the Cloverly Formation. These teeth are very peculiar but easy to identify because of their shape and fine bumpy texture.

Turtle taxa of the Cloverly Formation are *Naomichelys speciosa, Glyptops plicatulus,* and *Glyptops pervicax.* Fragments of turtle shell are often mistaken for dinosaur skin because of their texture. Turtle shell is usually about 0.5 centimeter (0.2 inch) thick, with a crinkled or bumpy texture.

little curvature
flattened ventral surface
— 5 cm (2 in.) —

Ornithomimus pes ungual

ventral view
flat surface

Ornithomimus pes phalanx

ventral view
flat bottom surface
lateral view

broken end
centrum
anterior
— 8 cm (3 in.) —

Ornithomimus caudal vertebra

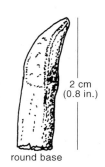
2 cm (0.8 in.)
round base
Titanosaur tooth

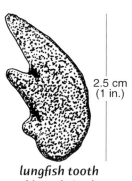
2.5 cm (1 in.)
lungfish tooth (dorsal view)

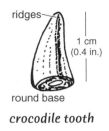

ridges

1 cm
(0.4 in.)

round base

crocodile tooth

Collectors commonly find the teeth and armor of crocodiles in the Cloverly Formation. A crocodile tooth is easy to identify because the base of its root is round and the tooth is conical. Crocodile armor plates, also known as scutes, have a distinctive central ridge and small holes all over the upper surface. The bottom surface is smooth. Most scutes are 3 to 4 centimeters (1.2 to 1.5 inches) in diameter.

Other fossils commonly found in the Cloverly Formation include rare mammal teeth, snails, and clams.

LATE CRETACEOUS
Two Medicine Formation
Campanian Stage
80 to 74 million years ago

The Two Medicine Formation is about 650 meters (about 2,100 feet) thick and represents the upland (western) equivalent of the Eagle and Judith River Formations. The formation consists primarily of greenish gray mudstone and thin beds of brown sandstone, but it also contains a few reddish beds near the top and rare coal beds near the bottom. Most sediments were deposited in streams, rivers, and lakes. In the southern region, near Wolf Creek, the Two Medicine Formation is loaded with volcanic rocks and is a distinctive green. The beach deposits of the Virgelle Sandstone lie below the Two Medicine Formation, and the black shales of the marine Bearpaw Formation lie above it.

MESOZOIC

Cretaceous

Two Medicine Formation

Jurassic

Triassic

Time position of the Two Medicine Formation

Ornithischian Dinosaurs

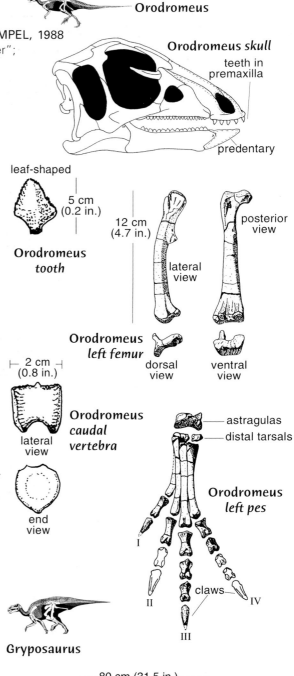

Orodromeus

Orodromeus makelai HORNER ET WEISHAMPEL, 1988

Orodromeus (ore-oh-DROH-me-us) = "mountain runner";
makelai (MAK-el-a-eye) = to honor Bob Makela
Type Specimen: MOR 294
(associated partial skeleton)
Discovered by: Bob Makela, 1983

Bob Makela found the first skeletons of *Orodromeus* associated with clutches of eggs at Egg Mountain, and researchers hypothesized that *Orodromeus* had laid the eggs. But, after several years, they discovered that the eggs actually belonged to a little carnivorous dinosaur called *Troodon*. The *Troodon* adults probably brought the *Orodromeus* to the nesting ground to feed to their babies.

Orodromeus was a small, 1.5-meter-long (about 5 feet), bipedal plant eater with tiny leaf-shaped teeth. Its skeletal elements were hollow and very lightly built, or gracile, indicating that it was a fast runner.

The vertebrae of *Orodromeus* are about the size of the end of a person's finger. An adult femur is about 15 centimeters (6 inches) long.

Orodromeus is common around Egg Mountain and other locations in the upper middle part of the formation, but it is extremely rare in other levels of the Two Medicine Formation. Collectors commonly find articulated specimens, but isolated elements have been found in microsites.

Hadrosauridae
(Duck-bill Dinosaurs)

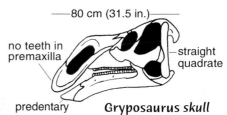

Gryposaurus

Gryposaurus latidens HORNER, 1992

Gryposaurus (GRIP-o-sore-us) = "hook-nosed reptile";
latidens (LAT-ih-dens) = "wide teeth"
Type Specimen: AMNH 5465
Discovered by: Tom Harwood, 1916
Referred Specimen: MOR 478 (bonebed)

Gryposaurus was a medium-size (8 to 9 meters or 26 to 29 feet long), bipedal plant eater that lived in large groups. One of three species of this

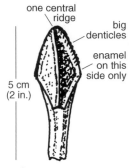

one central ridge
big denticles
enamel on this side only
5 cm (2 in.)

Gryposaurus dentary tooth

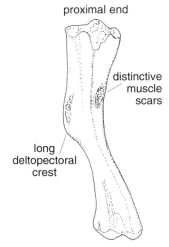

proximal end
distinctive muscle scars
long deltopectoral crest

Gryposaurus left humerus (posterior view)

Maiasaura peeblesorum

genus, *Gryposaurus latidens* was a hadrosaur with an arched nasal bone. The other two species of *Gryposaurus* are *G. notabilis* and *G. incurvimanus*. *Gryposaurus latidens* had very short, wide teeth in its jaws. Like all other duck-bills, *Gryposaurus* chewed its food.

It is difficult to distinguish the different kinds of duck-bill dinosaurs because most of the bones of different species look very similar and have received little study. The teeth of the dentary of *Gryposaurus latidens* are the only parts of the skeleton that one can readily identify without having a whole skeleton. The teeth of *Gryposaurus latidens* are identifiable because they have an enamel crown that is comparatively wider than the crowns of other hadrosaurian dinosaurs. Also, some teeth have two central ridges rather than one as in other hadrosaur teeth. The bones of the arms are very long compared to the bones of the hind legs, as is true with other primitive hadrosaurs, but this kind of comparison is only possible and useful if you find an associated skeleton.

Gryposaurus latidens is known only from Montana in the lower beds of the Two Medicine Formation. It is one of the older hadrosaurs from North America. Remains of this species have been found both as isolated articulated skeletons and disarticulated elements in bonebeds.

Maiasaura peeblesorum HORNER ET MAKELA, 1979
Maiasaura (MY-uh-sore-a) = "good mother reptile";
peeblesorum (pee-bulls-OR-um) = to honor the James and John Peebles families of Choteau, Montana
Type Specimen: YPM-PU 22405
Discovered by: Laurie Trexler, 1978
Referred Specimen: OTM f138

Maiasaura peeblesorum is the state fossil of Montana and has not yet been found anywhere outside the state. It is known only from the Two Medicine Formation, although it is closely related to *Brachylophosaurus* from the Judith River Formation. *Maiasaura* was a medium-size (8 meters or 26 feet long), bipedal, plant-eating hadrosaur.

Maiasaura was the first dinosaur associated with nests of babies. Paleontologists found baby maiasaurs in nests, and because different nests had different size babies, they hypothesized that the baby maiasaurs were living in the nest when they died. Researchers

Adult and nestling Maiasaura *skulls from the Two Medicine Formation of western Montana (NC). Nestling skull is 12 centimeters (about 5 inches) long.*

think that the parents probably cared for and brought food to the babies. *Maiasaura* skeletons of many sizes are also found in mass accumulations, suggesting that they may have traveled in giant herds when they were not nesting.

Like other duck-bill dinosaurs, *Maiasaura* is difficult to identify unless you have a good portion of the skull, especially the nasal crest region. Unlike that of other hadrosaurs, the *Maiasaura*'s nasal opening, or nose, is small and restricted to the front end of the snout. From the eyes to the nasal opening, the snout is long and lacks openings, similar to that of *Iguanodon*. The nasal crest of *Maiasaura* is very short and is concave on its anterior side.

Most of *Maiasaura*'s postcranial skeleton resembles that of *Brachylophosaurus* from the Judith River Formation. The undersides of the pes ungual phalanges of both *Maiasaura* and *Brachylophosaurus*, but of no other hadrosaurs, have a characteristic ridge.

The first specimens of *Maiasaura* were found near Egg Mountain west of Choteau in Teton County. Specimens have now been found in numerous other locations within the Two Medicine Formation of western Montana.

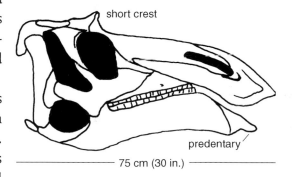

Maiasaura skull

short crest

predentary

75 cm (30 in.)

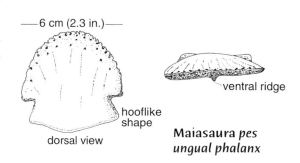

6 cm (2.3 in.)

dorsal view

hooflike shape

ventral ridge

Maiasaura *pes ungual phalanx*

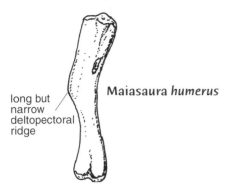

long but narrow deltopectoral ridge

Maiasaura *humerus*

Only two articulated skeletons of *Maiasaura* have been reported, but we can refer thousands of isolated bones to the species, including embryonic, juvenile, subadult, and adult bones. Eggs, coprolites, and footprints are also known for this taxon. Most *Maiasaura* specimens are found in bonebeds or on nesting horizons.

Prosaurolophus blackfeetensis

Prosaurolophus blackfeetensis

HORNER, 1992
Prosaurolophus (pro-sore-uh-LOW-fuss) = "before Saurolophus"; *blackfeetensis* (BLACK-feet-EN-sis) = to honor the people of the Blackfeet Nation
Type Specimen: MOR 454 (partial skeleton)
Discovered by: Jack Horner 1986

Prosaurolophus blackfeetensis is a medium-size (7 to 8 meters or 23 to 26 feet long), bipedal, plant-eating hadrosaur. Its remains are found only in the Two Medicine Formation of Montana, but a closely related animal named *Prosaurolophus maximus* is known from Alberta, Canada. Both species of *Prosaurolophus* have a characteristic small, scooped-out nasal crest just above the orbits.

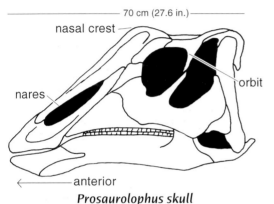

—— 70 cm (27.6 in.) ——
nasal crest
orbit
nares
← anterior
Prosaurolophus skull

Prosaurolophus blackfeetensis is the most common noncrested duck-bill from the upper strata of the Two Medicine Formation. Collectors have found it as associated skeletons or in bonebeds.

Hypacrosaurus stebingeri

HORNER ET CURRIE, 1994
Hypacrosaurus (hi-PACK-roe-sore-us) = "high-ridged reptile"; *stebingeri* (STEB-in-ger-i) = to honor Eugene Stebinger
Type Specimen: MOR 549 (articulated skeleton)
Discovered by: Jason Horner, 1988

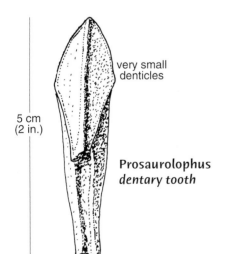

very small denticles

5 cm (2 in.)

Prosaurolophus dentary tooth

Like *Maiasaura*, *Hypacrosaurus stebingeri* is one of the most completely known dinosaurs in the world because there are remains of it from all stages of life, from embryo to adult. *Hypacrosaurus stebingeri* was a crested duck-bill dinosaur, or

Hypacrosaurus stebingeri

lambeosaur, that lived in western Montana when the Cretaceous Seaway was less than 100 miles from the Rocky Mountains. Studies of the skull of *Hypacrosaurus stebingeri* suggest that it may have been an intermediate evolutionary species between older and younger lambeosaur species. These crested hadrosaurs lived at the same time as the noncrested *Prosaurolophus blackfeetensis*.

Like all other crested hadrosaurs, *Hypacrosaurus stebingeri* has an expanded nasal crest on its head that it may have used as a resonating chamber for communication. Interestingly, the nasal crests of these crested dinosaurs didn't form until the animals had reached adult size. This suggests that the juveniles and subadults would have made a different sound than the adults did.

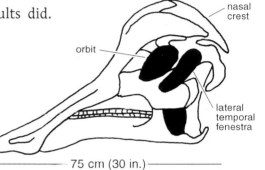

Hypacrosaurus skull

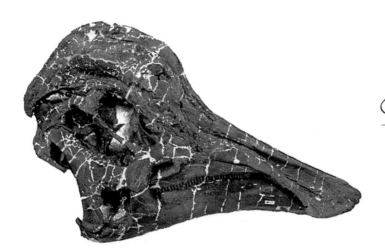

Skull of Hypacrosaurus stebingeri *from the Two Medicine Formation, Glacier County, Montana (BfN). Skull is 75 centimeters (30 inches) long.*

Young juvenile skull of Hypacrosaurus *from the Two Medicine Formation of northwestern Montana (BfN). Skull is 30 centimeters (12 inches) long.*

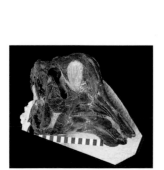

Skull of a juvenile Hypacrosaurus *from the Two Medicine Formation of western Montana (Pvt). Skull is 20 centimeters (about 8 inches) long.*

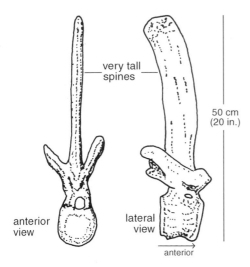

very tall spines

50 cm (20 in.)

anterior view

lateral view

anterior

Hypacrosaurus *dorsal vertebra*

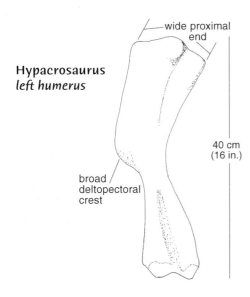

Hypacrosaurus *left humerus*

wide proximal end

40 cm (16 in.)

broad deltopectoral crest

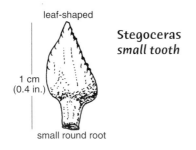

leaf-shaped

1 cm (0.4 in.)

Stegoceras *small tooth*

small round root

The only bones that paleontologists have identified as distinctive to *Hypacrosaurus stebingeri* are the thin bones that make up the hollow crest on its head, and these are usually crushed or broken when not attached to the skull.

In the postcrania, the vertebrae are distinctive in that they have extremely tall neural spines, but this is also true of *Hypacrosaurus* from the Edmonton Formation of Alberta, Canada. Because the two species of *Hypacrosaurus* are not found in the same formation, tall-spined vertebrae found in the Two Medicine Formation most likely belong to *Hypacrosaurus stebingeri*.

Specimens of *Hypacrosaurus stebingeri* are commonly found in massive bonebeds, suggesting that the animals may have traveled in large groups. Only one articulated skeleton of *Hypacrosaurus stebingeri* has been reported, but several massive bonebeds have yielded hundreds of adult and subadult skeletal parts. In addition, there are numerous eggs, embryos, and juveniles attributed to this taxon.

Pachycephalosauridae (Dome-headed Dinosaurs)

Stegoceras sp. (species undescribed)
Stegoceras (steg-AH-sir-us) = "covered horn"
Referred Specimen: MOR 480
(fronto-parietal dome)

Stegoceras is a pachycephalosaur, or bone-headed (dome-headed) dinosaur. These dinosaurs were gracile, or lightly built, bipedal plant eaters with small, leaf-shaped teeth. They were small, about 2 meters (6.5 feet) long. Some paleontologists think that the pachycephalosaurs head-butted like bighorn sheep do. But others, myself included, disagree, pointing out that their skeletons were lightly built and their skulls were not shaped right for head-butting and show no signs of trauma.

Stegoceras is rare, but every once in a while a person will pick up the thickened fronto-parietal part of a *Stegoceras* skull. To me, these domes look like a petrified meatball or a kneecap. On the underside, a small depression forms the top of the braincase.

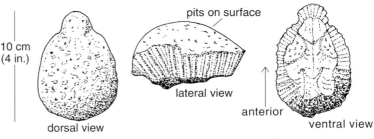

pits on surface

10 cm
(4 in.)

dorsal view

lateral view

anterior

ventral view

Stegoceras *fronto-parietal dome*

Not much is known about *Stegoceras* from the Two Medicine Formation because only three specimens have been found, and one of them is a juvenile. *Stegoceras* domes are generally found either in microsites or isolated.

Protoceratopsidae

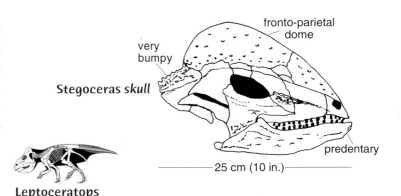

fronto-parietal dome

very bumpy

Stegoceras *skull*

predentary

25 cm (10 in.)

Leptoceratops

?*Leptoceratops* sp.

Leptoceratops (LEHP-to-sare-a-tops) = "slender horn face"
Referred Specimen: MOR 300 (articulated skeleton)

Protoceratopsian dinosaurs are small, primitive members of the ceratopsian family. Protoceratopsians have shields like the big ceratopsians do, but they don't have horns over their eyes or on their noses. They are generally uncommon, although a bonebed found recently in the Two Medicine Formation suggests these dinosaurs, like the larger ceratopsians, may have lived or traveled in groups.

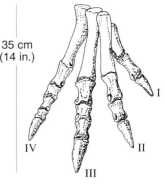

35 cm
(14 in.)

I

IV

II

III

Leptoceratops *right pes*

Two Medicine Formation protoceratopsian dinosaurs have not yet been described, and there are no other protoceratopsian dinosaurs named from Campanian-age sediments in North America. The one good specimen housed in the Museum of the Rockies collection bears a resemblance to *Leptoceratops,* known from Maastrichtian-age sediments. *Leptoceratops* was a very small (2 meters or about 6.5 feet long), quadrupedal plant eater. It was closely related to the Asian *Protoceratops* and more distantly related to the large horned dinosaurs such as *Triceratops.*

The teeth of protoceratopsians are identifiable because they have a single root, and the enameled part, or crown, has a large ridge. The teeth are about the size of the end of a grown person's

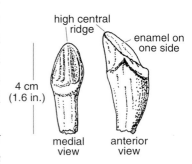

high central ridge

enamel on one side

4 cm
(1.6 in.)

medial view

anterior view

Leptoceratops *dentary tooth*

finger. Other horned dinosaur teeth have the large ridge but a double root.

Ceratopsidae

Styracosaurus ovatus

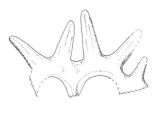

Styracosaurus
parietal frill

Styracosaurus ovatus GILMORE, 1917
Styracosaurus (stie-RACK-o-sore-us) = "spiked reptile";
ovatus (o-VAY-tus) = "egg-shaped"
Type Specimen: USNM 11869
Discovered by: George Sternberg, 1928

Styracosaurus ovatus is known only from some very fragmentary parts of one skull described by Charles Gilmore. It appears that it may have been closely related to *Styracosaurus albertensis* from Alberta, but with flattened spikes similar to *Einiosaurus*. More specimens of this dinosaur will have to be found before we will know whether this is a valid taxon or a variation of *Einiosaurus* or *Achelousaurus*.

Styracosaurus ovatus was probably about 7 meters (23 feet) long.

curved nasal horn

spike on
each side
of frill

———— 150 cm (59 in.) ————

Einiosaurus *skull*

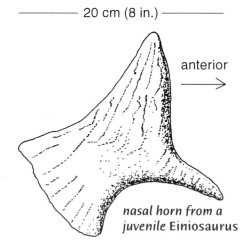

———— 20 cm (8 in.) ————

anterior
→

*nasal horn from a
juvenile* Einiosaurus

Einiosaurus procurvicornis SAMPSON, 1995
Einiosaurus (EEN-ee-o-sore-us) = "buffalo reptile";
procurvicornis (pro-CURVE-ih-CORN-iss) =
"forward curving horn"
Type Specimen: MOR 456-8-9-6-1 (skull)
Discovered by: Jack Horner
and Carrie Ancell, 1986

Einiosaurus was a medium-size (7 meters or 23 feet long), quadrupedal plant eater in the ceratopsian family. This dinosaur is particularly odd because its nose horn curved forward over the front of its face, kind of like a giant can opener. On the back of its frill, *Einiosaurus* had two long, straight spikes.

The nose horn is the only bone a person could find that would positively identify the animal as *Einiosaurus*. And the specimen would have to be an adult because the baby's horns look different than the adult's. The babies and juveniles of all the centrosaurine dinosaurs—horned dinosaurs with short squamosal bones of the skull—have similar skulls with no spikes on the frills and only short, straight nasal horns.

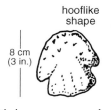

hooflike
shape

8 cm
(3 in.)

Einiosaurus *pes phalanx*

Skull of Einiosaurus *from the
Two Medicine Formation of
northwestern Montana (BfN).
Skull is 1.7 meters (5.5 feet) long.*

Einiosaurus remains are known from one extensive bonebed and a couple of isolated skeletons. They have been found only in Montana.

Achelousaurus horneri SAMPSON, 1995
Achelousaurus (ah-KEY-low-sore-us) = "achelous reptile"; *horneri* (horn-er-eye) = to honor John (Jack) Horner
Type Specimen: MOR 485 (skull)
Discovered by: Ray Rogers, 1987

This dinosaur is aptly named after Achelous, a river god from Greek mythology who transformed himself into a bull to fight Hercules. *Achelousaurus* was a medium-size (7 meters or 23 feet long), quadrupedal plant eater from the ceratopsian family. The skull of *Achelousaurus* is 1.5 meters (5 feet) long. Although *Achelousaurus* is technically a horned dinosaur, it didn't have horns. Rather, it had some deep gouges and rough ridges over its eyes and on its nose and two long spikes on the back of its frill. I think it's one of the weirdest dinosaurs ever (hmmm, it's named after me).

As with many other horned dinosaurs, the only part of *Achelousaurus* that is readily identifiable is a portion of its gnarly skull. A juvenile skull resembles that of a juvenile of any other centrosaurine because the horns, or gnarly knobs, and shields did not completely develop until the animals were full-grown.

Achelousaurus is represented by associated and articulated skeletons and elements from one bonebed in Montana.

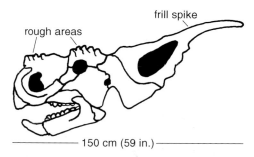

rough areas

frill spike

150 cm (59 in.)

Achelousaurus *skull*

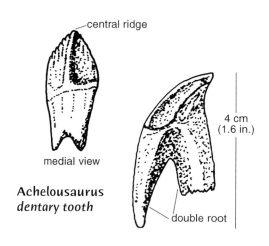

central ridge

medial view

4 cm
(1.6 in.)

**Achelousaurus
*dentary tooth***

double root

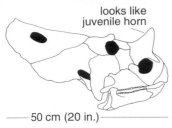

looks like
juvenile horn

— 50 cm (20 in.) —

skull of a juvenile
Brachyceratops

Brachyceratops montanensis GILMORE, 1914
Brachyceratops (brack-ee-SARE-uh-tops) = "short horned face";
montanensis (mon-TAN-en-sis) = in honor of Montana
Type Specimen: USNM 7951
Discovered by: Charles Gilmore, 1913

Brachyceratops montanensis is clearly a juvenile ceratopsian and most likely a juvenile of *Einiosaurus procurvicornis*, *Achelousaurus horneri*, or *Styracosaurus ovatus*, all of which are found in the area where Charles Gilmore collected this juvenile. All juvenile ceratopsians look alike, having small, pointed nasal horns even though the adults have no horns.

Charles Gilmore collected the only specimens referred to *Brachyceratops* from a single bonebed.

Ankylosauridae
Armored Dinosaurs

Euoplocephalus

Euoplocephalus sp. (species undescribed)
Euoplocephalus (u-oh-ploh-SEF-a-lus) = "well-protected head"
Referred Specimen: MOR 433 (partial skeleton)

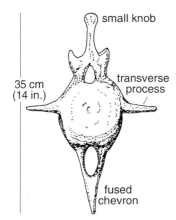

small knob

transverse
process

35 cm
(14 in.)

fused
chevron

Euoplocephalus
caudal vertebra

Euoplocephalus was a medium-size (5.5 meters or 18 feet long), quadrupedal plant eater with heavy armor on its body. Its tail ended in a large bony "club" of armored scutes that the animal used to defend itself. The tail vertebrae near the club are co-ossified, or fused, into a rigid rod that would have given the ankylosaur great leverage when swinging its club. Large bony plates that covered the ankylosaur's body would have protected its back from predatory bites. A heavy bony shield encased the ankylosaur's neck. The skull of *Euoplocephalus* is very solid, with armor plates welded, or fused, to the top.

The most common fossils of *Euoplocephalus* are its armor plates and its teeth. The teeth are small, leaf-shaped, and about the size of the end of a person's finger.

We don't know much about *Euoplocephalus* from the Two Medicine Formation because only one partial skeleton and two fragmentary skulls have been collected. Specimens of this taxon

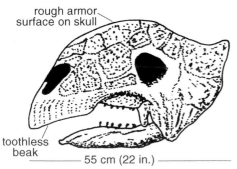

rough armor
surface on skull

toothless
beak

— 55 cm (22 in.) —

Euoplocephalus *skull*

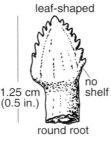

leaf-shaped

1.25 cm
(0.5 in.)

no
shelf

round root

Euoplocephalus
tooth

have been found as associated skeletons or more often as isolated teeth and armor plates in microsites in the Two Medicine Formation and in Alberta.

Edmontonia rugosidens

Edmontonia rugosidens GILMORE 1930
Edmontonia (ED-mon-TONE-ee-a) = named for the Edmonton Formation of Alberta; *rugosidens* (roo-GO-sih-dens) = "wrinkled hollow"
Type Specimen: USNM 11868
Referred Specimen: MOR 522 (partial skull)

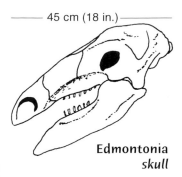

Edmontonia skull

Edmontonia was a medium-size (6 meters or 20 feet long), quadrupedal, plant-eating armored dinosaur. Unlike *Euoplocephalus*, *Edmontonia* didn't have a club on its tail. However, the skull of *Edmontonia* looks like a large, gnarly club with holes in it for eyes.

The leaf-shaped teeth of *Edmontonia,* about the size of the end of a person's finger, are relatively common, as are armor plates or fragments of plates. Armor plates can vary in size, and the largest is about the size of an adult person's hand. The armor plates of ankylosaurs are generally rather solid, whereas those of the nodosaurs are hollowed out on their undersides. Interestingly, the armor plates that have been found associated with the *Euoplocephalus* from the Two Medicine Formation have hollowed-out undersides like the armor of nodosaurs. We need to find more specimens of these taxa before we will resolve this mystery.

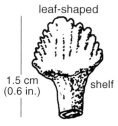

Edmontonia *tooth*

Associated skeletons have been collected, but this taxon is most commonly found as isolated teeth in microsites in Montana and Alberta.

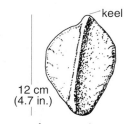

Edmontonia *armor plate*

Saurischian Dinosaurs

Daspletosaurus

Daspletosaurus sp. (species undescribed)
Daspletosaurus (das-PLEET-o-sore-us) = "frightful reptile"
Referred Specimen: MOR 590
(partial articulated skeleton)

Daspletosaurus is probably the most common tyrannosaur in the Two Medicine Formation, and it is most likely responsible for the isolated tyrannosaur teeth that collectors commonly find. *Daspletosaurus* resembled *Albertosaurus* but had differences within subtle structures of the cranial bones and in the number of tooth sockets. *Daspletosaurus* had fewer teeth than *Albertosaurus* did.

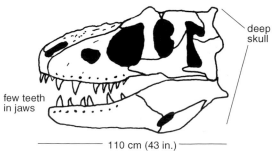

Daspletosaurus *skull*

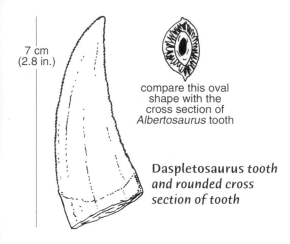

7 cm
(2.8 in.)

compare this oval
shape with the
cross section of
Albertosaurus tooth

Daspletosaurus *tooth
and rounded cross
section of tooth*

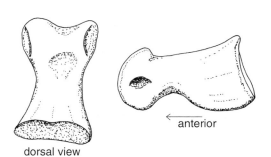

dorsal view

anterior

Daspletosaurus *pes phalanx*

The teeth of *Daspletosaurus* have about twenty serrations, or denticles, per centimeter, although it is not clear whether this characteristic is meaningful. In cross section, most of their teeth are more rounded than those of *Albertosaurus*. *Daspletosaurus* teeth are about the size of your thumb or slightly larger.

The phalanges are elongate and have deep depressions on their antero-lateral surfaces. The ungual phalanges have deep so-called blood grooves, and the claws are curved. Tyrannosaur bone has an extremely dense outer surface, so it typically looks shiny and smooth.

Skeletons of *Daspletosaurus* have been found both articulated and associated. Isolated teeth are common in microsites, and isolated, fragmentary limb bones are fairly common in Montana and Alberta.

*Daspletosaurus skull
from the Two Medicine
Formation, Glacier
County, Montana
(BfN). Skull is 1 meter
(about 3 feet) long.*

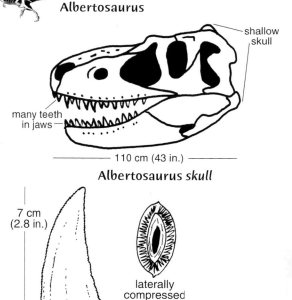

Albertosaurus

Albertosaurus skull

shallow skull

many teeth in jaws

110 cm (43 in.)

Albertosaurus sp. (species undescribed)
Albertosaurus (al-bert-o-SORE-us) = "Alberta reptile"

Collectors occasionally find teeth thought to belong to *Albertosaurus* in the Two Medicine Formation. These teeth have twenty-five serrations per centimeter, similar to serration numbers for *Albertosaurus* specimens from the Judith River Formation. The teeth have been found in microsites. I generally think of *Albertosaurus* teeth as being more compressed side-to-side, or laterally, and *Daspletosaurus* teeth as being more round in cross section, like those of a *Tyrannosaurus*—but I might be wrong.

7 cm (2.8 in.)

laterally compressed (narrow)

Albertosaurus tooth and cross section of tooth

Troodon formosus LEIDY, 1856
Troodon (TRUE-o-don) = "wound tooth";
formosus (for-MO-sus) = "pretty"
Type Specimen: ANSP 9259
Discovered by: Ferdinand V. Hayden, 1855
Referred Specimen: MOR 563 (skeleton)

Troodon was about 2 meters (6.5 feet) long and probably weighed around 100 pounds. The type specimen of *Troodon* is a single tooth collected by Ferdinand Hayden in 1855. Yet because these teeth have such large serrations, or denticles, with respect to the size of the tooth, they are easy to identify. In the Two Medicine Formation, *Troodon* teeth are fairly common, and some have been found in jaws associated with partial skeletons. Average teeth are about 1 centimeter (0.4 inch) long.

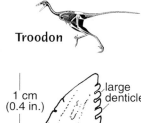

Troodon

Troodon has a very distinctive foot—one of its metatarsals is very reduced, apparently for carrying a sickle claw similar to that of the dromaeosaurids such as *Deinonychus*. The vertebrae of *Troodon* are hollow and shaped much like those of birds.

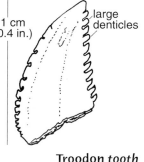

1 cm (0.4 in.)

large denticles

Troodon *tooth*

Remains of *Troodon* have been found in multispecies bonebeds and as associated and articulated skeletons. Teeth are common in microsites. *Troodon* eggs, some with embryos, are also known. The eggs of *Troodon* are 15 to 20 centimeters (6 to 8 inches) long. They have a blunt end and pointed end similar to bird eggs. The eggs have been found in arrangements suggesting that these dinosaurs nested in colonies on islands in lakes and along the shores of lakes. *Troodon* constructed mud-rimmed nests and sat

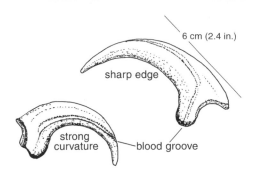

6 cm (2.4 in.)

sharp edge

strong curvature — blood groove

Saurornitholestes slashing claw from pes; Saurornitholestes manus claw

Saurornitholestes

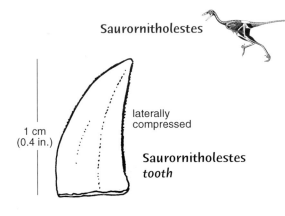

1 cm (0.4 in.)

laterally compressed

Saurornitholestes tooth

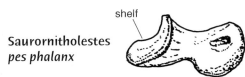

shelf

Saurornitholestes pes phalanx

on their eggs to incubate them, as modern birds do. Unlike birds, *Troodon* pushed their eggs partway into soil and arranged them in spirals. *Troodon* eggshell is smooth and only a couple of millimeters thick. When the babies hatched, they were active and could leave the nest, but they probably remained in the nesting area where the adults could protect them.

Saurornitholestes sp.
(species undescribed)
Saurornitholestes (sore-or-nith-o-LEST-ees) = "reptile bird robber"
Referred Specimen: MOR 660 (associated skeleton)

Saurornitholestes is a dromaeosaurid dinosaur closely related to *Deinonychus*, *Velociraptor*, and birds. It was a small dinosaur—only about 1.5 meters (5 feet) long and probably weighing not more than 13.5 kilograms (30 pounds)—but it was probably fierce and a fast runner. The sickle claws on each hind foot and the claws on the hands of *Saurornitholestes* were each extremely thin and recurved, suggesting that they were very

Saurornitholestes skeleton from the Two Medicine Formation of western Montana (BfN). Skeleton is 1.5 meters (5 feet) long.

sharp in life. All the bones of the skeleton were hollow, and the vertebrae were shaped like those of birds.

Two associated skeletons have been collected, and teeth are common in microsites. A typical tooth is anywhere from 5 to 10 millimeters (0.2 to 0.4 inch) long.

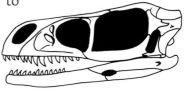

Bambiraptor skull

Bambiraptor feinbergi BURNHAM ET AL., 2000
Bambiraptor (BAM-bee-rap-tore) = "Bambi robber"; *feinbergi* (fine-BERG-eye) = to honor Michael and Ann Feinberg
Type Specimen: FIP 001 (The specimen is on display at the Florida Institute of Paleontology, but it is unclear whether the specimen is actually reposited there.)

Bambiraptor is a small maniraptoran theropod closely related to *Saurornitholestes*. It may actually be a juvenile *Saurornitholestes*. Its skull is elongated and has a large orbit and a large narial opening. Postcranial elements of this animal are very similar to those of *Saurornitholestes*.

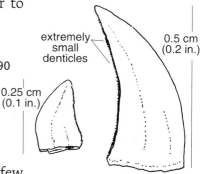

two variations of
Richardoestesia *teeth*

Richardoestesia gilmorei CURRIE, RIGBY ET SLOAN, 1990
Richardoestesia (ri-card-o-EES-zha) = honoring Richard Estes; *gilmorei* (GILL-more-i) = honoring Charles Gilmore
Type Specimen: NMC 343
Discovered by: Charles Gilmore, 1924
Referred Specimens: TM-088 (microsite)

Richardoestesia is known from a set of partial jaws and a few unerupted teeth from the Dinosaur Park Formation of Alberta. Teeth of similar morphology are known from the Two Medicine Formation and are distinctive. They are small, have serrations restricted mainly to their posterior edges, and have extremely small denticles, up to 12 per millimeter (305 per inch). Also, many teeth assigned to this genus are not curved as most other theropod teeth are. Most of the teeth are between 2 to 4 millimeters (about 0.1 inch) long.

Struthiomimus sp. (species unidentified)
Struthiomimus (strooth-ee-o-MIME-us) = "ostrich imitator"
Referred Specimen: MOR 450 (articulated foot)

Struthiomimus

All the postcranial bones of the ornithomimosaurs, or "ostrich dinosaurs," resemble one another, so it is difficult to identify the bones of the feet, which are the most common elements found

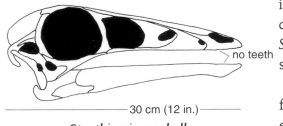

no teeth

— 30 cm (12 in.) —

Struthiomimus skull

**Struthiomimus
tibia**

bone is very
hollow like
that of
a bird

45 cm
(18 in.)

astragulas

in most formations. Paleontologists refer the ornithomimosaur from the Two Medicine to *Struthiomimus* because the specimens are from sediments of similar age.

Toe bones, or phalanges, are occasionally found in microsites, but most other elements are either rare or have been misidentified. The tibia is relatively easy to identify because it is long, straight, and narrow. Like other theropod bone, the surface of all ornithomimid bone is very dense and shiny.

Refer to the ornithomimids from other formations to see additional bone drawings.

Unidentified Theropod Teeth

The Two Medicine Formation does not have a lot of microsites, but at those it does have, researchers have collected teeth from numerous small, unidentified theropods, including the toothed birds. Many of those teeth may well represent new taxa, but others may represent teeth of known animals for which we lack complete dentitions. Teeth seldom have distinctive features that allow us to name new animals on their basis. Although Edward D. Cope, in the late 1800s, named a diversity of dinosaurs on the basis of teeth, none of those taxa currently stand as valid taxa. *Troodon formosus*, which Leidy originally described in 1856, remains one of the few taxa that was named based on a single tooth—and that is only because the tooth is so unusual.

Common Nondinosaurian Vertebrate Fossils from the Two Medicine Formation

Vertebrate remains representing animals that lived in streams and rivers are uncommon in the lower strata, nearly absent in upper-middle beds, and relatively common in the upper strata. Microsites are equally dispersed. In the upper-middle beds, where fluvial faunas are absent, terrestrial forms are common.

In the lower and upper strata, garfish (*Lepisosteus*) scales are the most common remains of fishes. The flattened teeth of the mollusk-eating ray *Myledaphus* are occasionally found in microsite deposits.

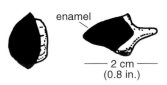

enamel

— 2 cm —
(0.8 in.)

garfish scales

In the upper strata, amphibian remains are known but extremely rare and probably represent the salamander *Scapherpeton*.

In the upper-middle strata of the Two Medicine Formation, the remains of the nondinosaurian teiid lizards and varanoid lizards are relatively common, usually closely associated with the dinosaurs *Maiasaura* and *Orodromeus*. Hard caliche limestones typically encase these remains, however, making them very difficult to find. Turtle remains are rare in these strata, and tortoises are the most commonly represented. Champsosaur and crocodilian remains are very rare in the upper-middle strata, though they are occasionally found in lag deposits in channel sandstones.

The most common turtle from the lower strata is *Compsemys*, identified by the small raised knobs or denticles that cover the surface of its carapace. In the upper strata, trionychids and baenids are common. Whole or partial carapaces are not uncommon in the upper strata, although cranial material is very rare.

In the upper strata, the remains of champsosaurs and crocodilians are also fairly common, usually associated with microsites.

Mammal remains are rare in the Two Medicine Formation, probably because microsites are not common. Most of the mammal remains have been found on the Egg Mountain nesting sites in the upper-middle strata. Both a small plant eater called a multituberculate *(Cimexomys)* and a marsupial *(Alphadon)* have been identified on the basis of teeth, but they are extremely small and require a hand lens to be seen.

Time position of the
Judith River Formation

LATE CRETACEOUS
Judith River Formation
Campanian Stage
78 to 74 million years ago

The Judith River Formation was deposited in a lowland environment between the Two Medicine Formation to the west and the inland sea to the east. The formation is generally tan and consists mainly of channel sandstones and siltstones with minor beds of

MESOZOIC	Cretaceous	Judith River Formation
	Jurassic	
	Triassic	

floodplain mudstones. The layers of massive sandstone apparently represent relatively large meandering rivers. The formation contains occasional thin beds of coal, which indicates that there were some areas of standing water and local swamps. The Judith River Formation measures about 152 meters (500 feet) thick near its western edge near the east end of the Sweetgrass Hills. It thins to less than 3 meters (10 feet) thick at its eastern edge near Saco. The formation is bound by the marine Claggett Formation beneath and the marine Bearpaw Formation above.

Dinosaur remains in the Judith River Formation are typically fragmentary, and as a result, very few dinosaur taxa have been positively identified. Collectors have reported only four articulated dinosaur skeletons.

The best place to see the Judith River Formation is around the city of Havre. Fossil hunters must have permits to collect on all state and federal lands and obtain permission from all private landowners.

Ornithischian Dinosaurs

An undescribed "hypsilophodontid" has been found in the Judith River Formation but has not yet been described. It appears to have been an animal about 1 meter (3 feet) long, and probably looked similar to *Orodromeus makelai* from the Two Medicine Formation at Egg Mountain. The single, associated specimen was found in a *Brachylophosaurus* bonebed.

Stegoceras validus LAMBE, 1902
Stegoceras (ste-GAH-sir-us) = "covered horn";
validus (VAL-ih-dus) = "strong"
Type Specimen: NMC 515
Referred Specimen: MOR 391
(fronto-parietal dome)

As is true in all Cretaceous formations, pachycephalosaurid dinosaurs are very rare in the Judith River Formation. Their identifiable remains are primarily the thick bony skull cap, called the fronto-parietal dome, that characterizes the taxon. Specimens of *Stegoceras* from the Judith River Formation usually consist of either the fronto-parietal dome or some other, smaller pieces of the skull. Very little postcranial skeletal material has been positively identified. An average *Stegoceras* dome is about the size of a man's fist.

Stegoceras was only about 1.5 meters (5 feet) long and probably weighed no more than 31 kilograms (about 70 pounds). It's hard to imagine these little, lightly built dinosaurs crashing their domed heads together, as several paleontologists have proposed. I think it's more likely that the dome functioned simply as a display organ for attracting mates.

Stegoceras domes and other cranial elements are generally found in microsites or as isolated bones.

Gryposaurus incurvimanus

PARKS, 1920
Gryposaurus (GRIP-o-sore-us) = "hook-nosed reptile"; *incurvimanus* (in-curve-ih-MAN-us) = "curved hand"
Type Specimen: ROM 4514
Referred Specimen: UM-5204 (partial skull)

The name *Gryposaurus* has a complicated history. It was first coined by Lambe in 1914 (*Gryposaurus notabilis*) for a hadrosaur from the Campanian of Alberta. Then in 1942, Lull and Wright merged *Gryposaurus* and *Kritosaurus,* a taxon from New Mexico, thinking them to be the same. *Kritosaurus* had been named in 1910, and therefore had priority. In 1992, I reexamined these dinosaurs and separated them back into two separate genera. I think *Gryposaurus* is significantly

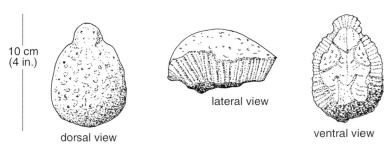

10 cm (4 in.)

dorsal view

lateral view

ventral view

Stegoceras *frontal parietal dome*

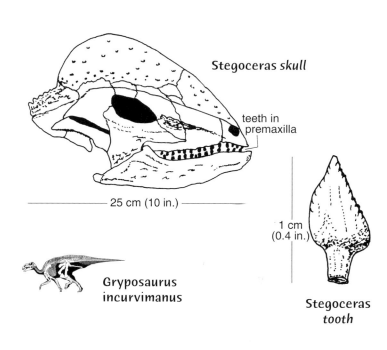

Stegoceras *skull*

teeth in premaxilla

25 cm (10 in.)

1 cm (0.4 in.)

Gryposaurus incurvimanus

Stegoceras *tooth*

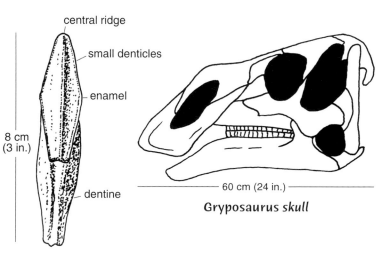

central ridge

small denticles

enamel

8 cm (3 in.)

dentine

Gryposaurus *tooth (medial view)*

60 cm (24 in.)

Gryposaurus *skull*

different from *Kritosaurus* even though their nasal bones look similar.

An average adult *Gryposaurus* was about 8 meters (25 feet) long and weighed around 2.5 tons. The nasal bone of *Gryposaurus* is the animal's most distinct element, and it looks like a large, bent hook. *Gryposaurus* teeth are broad for their height compared with those of other taxa of hadrosaurs.

Gryposaurus remains are relatively common in the upper strata of the Judith River Formation. They are generally found either as associated skeletons or, more commonly, as isolated bones.

Brachylophosaurus canadensis STERNBERG, 1953
Brachylophosaurus (brach-ee-LOW-foh-sore-us) = "short-crested reptile";
canadensis (can-ah-DEN-sis) = to honor Canada
Type Specimen: NMC 8893
Discovered by: Charles M. Sternberg, 1936

Brachylophosaurus goodwini HORNER, 1988
Brachylophosaurus (brack-ee-LOW-foh-sore-us) = "short-crested reptile";
goodwini (GOOD-win-eye) = to honor Mark Goodwin
Type Specimen: UCMP 130139
Discovered by: Mark Goodwin, 1986
Merged with *Brachylophosaurus canadensis*
by Prieto-Marquez, 2001

Skull of Brachylophosaurus from the Judith River Formation of northeastern Montana (BLM). Skull is 80 centimeters (31 inches) long.

Brachylophosaurus is the most common hadrosaur found in the lower strata of the Judith River Formation. An average adult was about 7 meters (23 feet) long and weighed between 2 and 3 tons. Several of its skull bones are distinctive, although it is difficult to distinguish some of the facial bones from those of the closely related *Maiasaura*. *Brachylophosaurus* has a broad, low, solid crest on the top of its skull that in adults extends posteriorly, or backward, over the entire back half of the skull.

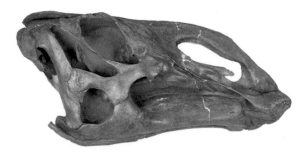

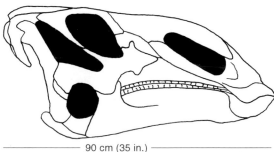

———— 90 cm (35 in.) ————

Brachylophosaurus skull

The postcranial skeleton of *Brachylophosaurus* has many characteristics similar to crested dinosaurs such as *Hypacrosaurus*. The neural spines of the posterior dorsal vertebrae and anterior caudal vertebrae are very tall, and the forearm (ulna and radius) is longer than the humerus.

Brachylophosaurus canadensis is known from articulated skeletons and bonebed occurrences and is most common in northeastern Montana.

Avaceratops lammersi

DODSON, 1986
Avaceratops (ay-va-SER-ah-tops) = "Ava's horn face" (for Ava Cole); *lammersi* (LAM-erz-i) = to honor the Lammers family of Shawmut, Montana
Type Specimen: ANSP 15800
Discovered by:
Eddy Cole, 1981
Referred Specimen:
MOR 692 (skull)

Cranial specimens of *Avaceratops* are not complete, and yet it is the best known ceratopsian from the Judith River Formation of Montana. Peter Dodson, who has studied this dinosaur in detail, thinks *Avaceratops* might be closely related to *Triceratops* because it appears to have lacked any openings in its neck shield. At the Museum of the Rockies, we have a skull found in the Judith River Formation that lacks holes in its parietal shield, has

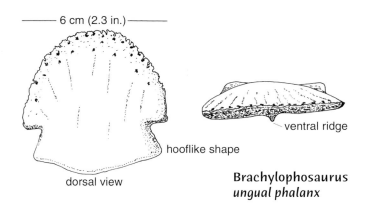

— 6 cm (2.3 in.) —

hooflike shape

dorsal view

ventral ridge

Brachylophosaurus ungual phalanx

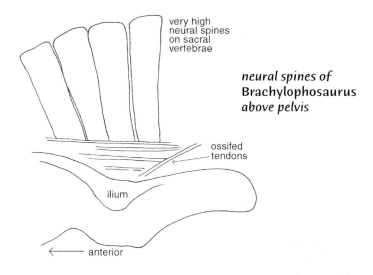

very high neural spines on sacral vertebrae

neural spines of Brachylophosaurus above pelvis

ossifed tendons

ilium

← anterior

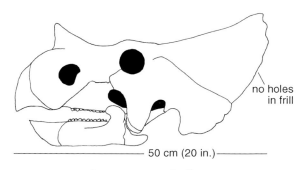

no holes in frill

— 50 cm (20 in.) —

Avaceratops *skull*

large horns over its eyes, and may well be an adult *Avaceratops*. However, as is so often the case, we need more specimens before we will know for sure what this dinosaur looked like and who its relations were.

Saurischian Dinosaurs

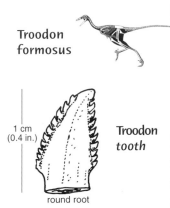

Troodon
formosus

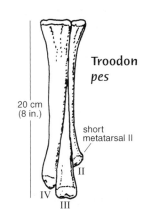

1 cm
(0.4 in.)

Troodon
tooth

round root

Troodon
pes

20 cm
(8 in.)

short
metatarsal II

II

IV

III

Troodon formosus LEIDY, 1856
Troodon (TRUE-o-don) = "wound tooth";
formosus (for-MO-sus) = "pretty"
Type Specimen: ANSP 9259
Discovered by: Ferdinand V. Hayden, 1855
Referred Specimen: JR-144Q (tooth)

Troodon specimens from the Judith River Formation consist of teeth. Because the teeth have large serrations, they are relatively easy to identify. Most *Troodon* teeth are no more than 1 centimeter (0.4 inch) long. Unlike other theropod teeth, the teeth of *Troodon* have round roots, like primitive ornithischian teeth do. Some researchers have mistaken *Troodon* teeth with roots for the teeth of animals such as *Stegoceras* and *Orodromeus*.

Troodon teeth are commonly found in microsites.

Albertosaurus libratus LAMBE, 1914
Albertosaurus (al-bert-o-SORE-us) = "Alberta reptile";
libratus (lib-RA-tus) = "balance"
Type Specimen: NMC 2120
Discovered by: Alberta Geological Survey
Referred Specimen: MOR 657 (partial skeleton)

According to paleontologist Dale Russell, two tyrannosaur taxa, *Albertosaurus* and *Daspletosaurus*, are known from Campanian age sediments in Canada. Presumably these animals also lived in Montana. There are few differences between these two animals. Possibly the biggest difference is that *Daspletosaurus* has a deeper skull and slightly fewer teeth in its maxilla than *Albertosaurus* does. Unfortunately, good cranial material is rare from this formation in Montana, and it's difficult to determine which taxa might be present. Judging from a couple of maxillae, one associated with a partial skeleton, *Albertosaurus* seems to be the best candidate for the Montana specimens.

The teeth, without their roots, can be as long as 12 centimeters (4.75 inches). The serrations on these teeth are very small, with about twenty-five serrations per centimeter (sixty-three serrations per inch).

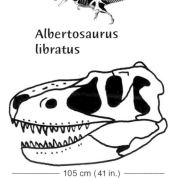

Albertosaurus
libratus

105 cm (41 in.)

Albertosaurus *skull*

Albertosaurus remains are usually represented by isolated teeth, mainly in microsites, or as associated and articulated skeletons.

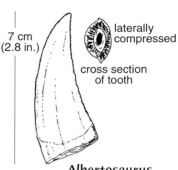

7 cm
(2.8 in.)

laterally compressed

cross section
of tooth

Albertosaurus
tooth

Dromaeosaurus albertensis MATTHEW ET BROWN, 1922

Dromaeosaurus (DROH-may-o-sore-us) = "running reptile";
albertensis (AL-bert-EN-sis) = for the province of Alberta
Type Specimen: AMNH 5356
Referred Specimen: JR-144Q (tooth)

Skeletal remains of *Dromaeosaurus*, other than its teeth, are probably difficult to distinguish from either juvenile tyrannosaurs or other dromaeosaurs. You can identify the teeth of *Dromaeosaurus*, however, on the basis of the position of their serrations. On teeth from both upper and lower jaws, the anterior serrations are on the medial side rather than the center. At the points of the teeth, the serrations are oriented anterior and posterior, meaning forward and backward. But as the serrations drop toward the gum line, they twist to the lingual side, or toward the tongue.

Dromaeosaurus teeth are usually no more than 5 centimeters (2 inches) long and are usually found in microsites.

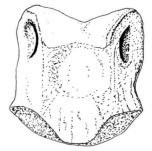

Albertosaurus *pes phalanx IV₃ (the third phalanx on the fourth digit) in dorsal view*

Aves

Bird remains are rare in the Judith River Formation, and they are often difficult to distinguish from remains of small theropods. However, the tarsometatarsus—which includes the three metatarsals plus the distal tarsal, or ankle, elements—is unmistakable. The metatarsals are clearly fused together to form a single unit. These fossils are about 5 centimeters (2 inches) long.

10 cm (4 in.)

Albertosaurus *pes phalanx ungual III₄ (the fourth phalanx on the third digit)*

Unidentified Theropod Teeth

Microsites abound in the Judith River Formation of Montana. As a result, collectors have found numerous small theropod teeth that are not assignable to any previously described animals. Some of these teeth might be referable to *Richardoestesia*, but others are simply unknown. "*Paronychodon*"-like teeth may be abnormal teeth of known taxa such as *Saurornitholestes* and *Troodon*.

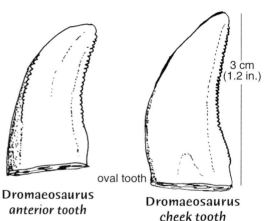

3 cm
(1.2 in.)

oval tooth

Dromaeosaurus
anterior tooth

Dromaeosaurus
cheek tooth

Common Nondinosaurian Vertebrate Fossils from the Judith River Formation

The microsites common in most of the Judith River Formation yield numerous nondinosaurian vertebrate remains. Most of the microsites lie at the interface between channel sandstones and overbank mudstones. Screen-washing is the best method of retrieving these kinds of specimens.

Both cartilaginous and bony fish remains are common in the Judith River Formation. Nearshore marine deposits commonly yield typical shark teeth that average about 2 centimeters (0.8 inch) long. The flattened, hexagonal teeth of the ray *Myledaphus* are also common, as are teeth of the sawfish *Ischyrhiza*.

Bony fishes include *Lepisosteus* (garpike), represented by abundant scales and occasional skull plates and vertebrae; *Acipenser* (sturgeon), usually represented by fin spines; *Amia* (bowfin), most commonly represented by vertebrae and jaw and gill plates; the mollusk-eating fish *Paralbula*, represented by small, domed-shape teeth; and *Belonostomus,* represented by jaw fragments and probably by vertebrae.

The most common amphibian remains are referable to salamanders and consist of vertebrae and jaw fragments. Two taxa, including *Scapherpeton* and *Prodesmodon,* have been identified.

Lizard remains are very rare in the Judith River Formation, but turtles, champsosaurs, and crocodilians are common. The most common turtles are the trionychids, or soft-shell turtles, such as *Aspideretes* and *Basilemys*. Baenid turtles such as *Baena* and *Boremys* are also common. Most turtle remains consist of chunks of the top shells (carapaces), bottom shells (plasterons), or phalanges.

Most microsites commonly yield the remains of *Champsosaurus* in the form of vertebral centra, ribs, and limb ends. Cranial elements are rare. Individual centra of champsosaurs have a distinctive hourglass pattern on their dorsal surfaces. The centra also usually have short, transverse processes.

Most mammal remains from the Judith River Formation consist of isolated teeth or jaw fragments with one or two teeth. The most common and identifiable mammalian teeth are those belonging to multituberculates. These small mammals were apparently plant eaters, because they have short grinding molars and bladelike, slicing premolars.

6 sides
dorsal view

└─5 cm─┘
(0.2 in.)

Myledaphus fish tooth (two views)

2 cm (0.8 in.)

garfish scale

0.25 cm (0.1 in.)

Paralbula fish tooth

LATE CRETACEOUS
Livingston Group—Miners Creek, Billman Creek, and Hoppers Formations
Maastrichtian Stage
68 to 65 million years ago

The Miners Creek, Billman Creek, and Hoppers Formations are part of the Livingston Group, sediments predominantly from volcanic rocks. Streams and rivers deposited the Miners Creek, Billman Creek, and Hoppers Formations in a basin east of the Bridger Range during the initial upheaval of the mountain range. The rocks of these formations are very hard, dense, and difficult to dig. As a result, very few dinosaur remains have been reported in these rocks. The few dinosaur fossils that collectors have found are black and typically fragmentary. These formations are in need of systematic collection.

MESOZOIC	Cretaceous	Livingston Group
	Jurassic	
	Triassic	

Time position of the Livingston Group

The Miners Creek, Billman Creek, and Hoppers Formations are bound by the beach deposits of the Lennep Sandstone beneath and the Paleocene Fort Union Group above. All three of these formations taken together are equivalent to the St. Mary River Formation to the west and north, and the Hell Creek Formation to the east.

The best places to see the rocks of the Miners Creek, Billman Creek, and Hoppers Formations are in roadcuts around the town of Livingston.

Tyrannosauridae (genus and species unidentified)
Referred Specimen: MOR 002

A partial skeleton and a few teeth have been found and identified as a tyrannosaurid, but so far the genus has not been determined. The teeth are laterally compressed, similar to the teeth of *Daspletosaurus*. These might also represent young *Tyrannosaurus rex* teeth.

Hadrosaurinae (genus and species unidentified)
Reference Specimen: MOR 400

Various fragments of hadrosaurs have been collected from a number of localities, but few are identifiable beyond the family Hadrosauridae. A partial ischium is referred to the Hadrosaurinae and represents a hadrosaur without a hollow crest, such as *Edmontosaurus*.

LATE CRETACEOUS
St. Mary River Formation (Including the Willow Creek Formation) Maastrichtian Stage
72 to 65 million years ago

The St. Mary River Formation rests on the Two Medicine Formation to the south and west and on the marine Bearpaw Formation to the north and east. This formation consists primarily of greenish gray sandstones, siltstones, and mudstones. Small streams that flowed westward out of the young Rocky Mountains deposited most of these sediments. In some areas, particularly in the north, the red Willow Creek Formation overlies the St. Mary River Formation. The Willow Creek Formation apparently represents desert conditions. Somewhere within this red unit is the Cretaceous/Tertiary boundary, but we don't have a good understanding of the geology of this area. The St. Mary River and Willow Creek Formations overlie the Bearpaw Formation and underlie unidentified Tertiary rocks.

Around the town of Augusta, in the southern extent of the St. Mary River Formation's exposure, thrust faulting of the Rocky Mountains has highly disturbed much of the formation. As a result the beds typically stand on end. You can best see the St. Mary River Formation on the west side of U.S. 89 at the Two Medicine River bridge on the Blackfeet Reservation. Look for the Willow Creek Formation near Duck Lake east of Babb. A permit is required for collecting on the reservation.

Time position of the St. Mary River Formation

MESOZOIC	Cretaceous	St. Mary River Formation
	Jurassic	
	Triassic	

Montanaceratops cerorhynchus

BROWN ET SCHLAKJER, 1942
Montanaceratops (mon-TAN-a-SARE-a-tops) = "Montana horn face";
cerorhyncus (ser-o-RINK-us) = "horned nose"
Type Specimen: AMNH 5464
Discovered by: Barnum Brown, 1916
Referred Specimen: MOR 542 (partial skeleton)

Montanaceratops is a very interesting dinosaur because researchers originally described it as having a nasal horn, and recent analyses indicate that the horn was misidentified. New studies will help us determine its relationships with the other horned dinosaurs.

As it stands now, *Montanaceratops* is the only protoceratopsian dinosaur known from the St. Mary River Formation, although it may turn out to be the same genus as *Leptoceratops* from rocks of a similar age in Alberta and Wyoming.

Additional Dinosaur Taxa

We know a number of other dinosaur taxa from the St. Mary River Formation that have not yet been described. Among these are a tyrannosaurid, a lambeosaurine hadrosaur, and a ceratopsian. The lambeosaur likely is closely related to *Hypacrosaurus altispinus*, and the ceratopsian probably is related to *Pachyrhinosaurus*. *Saurolophus* may also be present in this srtata.

Nondinosaurian Vertebrate Fossils from the St. Mary River Formation

Little is known of the nondinosaurian vertebrate fauna of the St. Mary River Formation. A turtle carapace and some crocodilian material have been collected.

LATE CRETACEOUS
Hell Creek Formation
Maastrichtian Stage
70 to 65 million years ago

The Hell Creek Formation is one of the most extensively exposed Mesozoic formations in the state, cropping out over a vast area of eastern Montana. The formation consists mainly of tan sandstones, siltstones, and mudstones with

Time position of the Hell Creek Formation

MESOZOIC	Cretaceous	Hell Creek Formation
	Jurassic	
	Triassic	

occasional thin beds of coal or peat. The formation represents a lowland environment deposited during the retreat of the inland seaway near the end of Cretaceous time.

Under the Hell Creek Formation lie the beach deposits of the Fox Hills Sandstone and over it lie terrestrial sediments of the Paleocene Tullock Formation. The boundary between the Tullock and Hell Creek Formations is the Cretaceous/Tertiary boundary.

Thescelosaurus
neglectus

long tail

Ornithischian Dinosaurs
"Hypsilophodontid" Grade

Thescelosaurus neglectus GILMORE, 1913
Thescelosaurus (THES-cell-o-sore-us) = "marvelous reptile"; *neglectus* (neg-LECK-tus) = "overlooked"

Type Specimen: USNM 7757
Referred Specimen: MOR 1106 (partial skeleton)

"Hypsilophodontid" grade dinosaurs are relatively rare in the fossil record, probably because most of them are small and have hollow bones. *Thescelosaurus* from the Hell Creek Formation is not small, however. *Thescelosaurus* was probably 3 to 4 meters (10 to 13 feet) long and may have weighed 225 kilograms (500 pounds). Its bones seem to be relatively common, even though articulated and associated skeletons are extremely rare. Most of its bones look somewhat similar to those of hadrosaurs, except generally smaller. The most commonly found and identifiable elements of *Thescelosaurus* are its toe bones, or phalanges. These are characteristically compressed and resemble the toe bones of modern camels. *Thescelosaurus* had clawed feet. You can also recognize the vertebrae by the rough ridges and grooves along the sides of the centra.

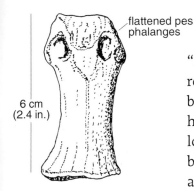

flattened pes
phalanges

6 cm
(2.4 in.)

Thescelosaurus
flattened pes
phalanges (dorsal view)

Thescelosaurus
vertebral centra

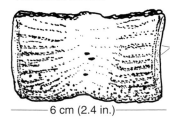

distinct
ridges
on sides

6 cm (2.4 in.)

Buganosaura infernalis GALTON, 1995
Buganosaura (boo-GAN-o-sore-a) = "large feminine cheek reptile"; *infernalis* (in-fern-AL-iss) = "the underworld"

Type Specimen: SDSM 7210
Referred Specimen: MOR 979
(articulated skeleton)

Buganosaura so closely resembles *Thescelosaurus* that we can't clearly distinguish their postcranial skeletons from one another. Among other relatively subtle features, the most significant cranial difference is in the shape of the maxilla and

Buganosaura
infernalis

whether it has an extended lateral ridge. The two animals may also differ slightly in overall size, and one or both of the taxa might have had small dermal ossicles, small bony armor embedded in their skin.

Buganosaura is known from a complete skull and skeleton collected by the Museum of the Rockies.

The teeth of *Buganosaura* have a very primitive leaflike shape and are about the size of the end of a person's little finger.

Like *Thescelosaurus*, *Buganosaura* was 3 or 4 meters (10 to 13 feet) long and may have weighed 225 kilograms (500 pounds) as an adult.

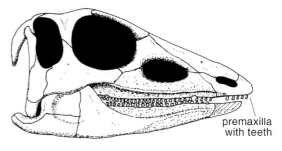

— 30 cm (12 in.) —

Buganosaura *skull*

Pachycephalosauridae

Pachycephalosaurus wyomingensis
GILMORE, 1931
Pachycephalosaurus (pack-ee-SEF-a-low-SORE-us) = "thick-headed reptile"; *wyomingensis* (wie-OHM-ing-EN-sis) = to honor the state of Wyoming
Type Specimen: USNM 12031
Referred Specimen: MOR 1100 (fronto-parietal dome)

Like the fronto-parietal domes of *Stegoceras*, the thickened cranium of *Pachycephalosaurus* is the most commonly found and identifiable part of this taxon. The skull of *P. wyomingensis* can measure more than 60 centimeters (23 inches) long and has a solid dome from 10 to 13 centimeters (4 to 5 inches) thick. Some bony bumps rise on the sides of the dome and in some specimens extend down the animal's nose.

Pachycephalosaurus may have been between 4 and 5 meters (13 to 16 feet) long. Skull remains are generally found isolated.

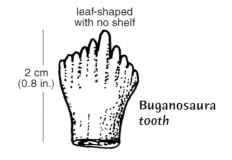

Skull of Buganosaura *from the Hell Creek Formation, Dawson County, Montana (MT). Skull is 20 centimeters (8 inches) long.*

leaf-shaped with no shelf

2 cm (0.8 in.)

Buganosaura *tooth*

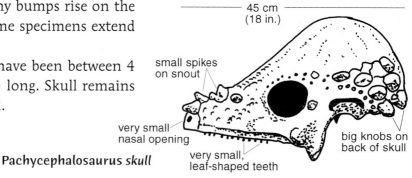

— 45 cm (18 in.) —

small spikes on snout

very small nasal opening

very small, leaf-shaped teeth

big knobs on back of skull

Pachycephalosaurus *skull*

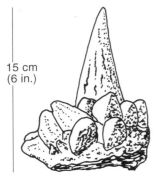

15 cm
(6 in.)

*spike from the back of
a Stygimoloch skull*

Stygimoloch spinifer GALTON ET SUES, 1983
Stygimoloch (stig-ee-MOH-lock) = "fierce river of the underworld";
spinifer (SPIN-i-fur) = "carrying thorns"
Type Specimen: UCMP 119433

Stygimoloch has a fronto-parietal dome similar to that of *Pachycephalosaurus*, but with additional, larger and longer bony horns or spikes on the back side of its dome. These bony spikes can be 10 centimeters (4 inches) long. The spikes are probably the most identifiable parts of this animal. Postcranial bones of *Stygimoloch* closely resemble the bones of *Pachycephalosaurus* and *Thescelosaurus*. Unusually thickened ossified tendons are known to be from *Stygimoloch*.

Ceratopsidae

Triceratops

Triceratops prorsus MARSH, 1890
Triceratops (try-SARE-a-tops) = "three horned face";
prorsus (PROR-sus) = "forward"
Type Specimen: YPM 1822
Referred Specimen: MOR 004 (skull)

According to Cathy Forster, who has studied *Triceratops*, there are two species of this taxon. *Triceratops prorsus* has a long nasal horn, short snout, and a relatively shortened frill. All the other parts of the two species of *Triceratops* are probably similar. Collectors have found so few skeletons of *Triceratops* that it is not clear how much variation exists between species, or even between *Triceratops* and *Torosaurus*.

*Triceratops prorsus skull
from the Hell Creek
Formation, Garfield County,
Montana (USFW). Skull is
2.25 meters (7.4 feet) long.*

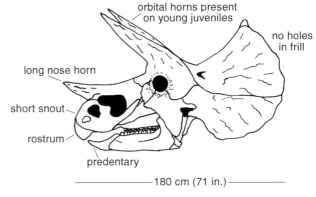

orbital horns present
on young juveniles

no holes
in frill

long nose horn

short snout

rostrum

predentary

——————— 180 cm (71 in.) ———————

Triceratops prorsus *skull*

Occasionally collectors can find complete nasal horns of young *Triceratops* that apparently fell off the skull during desiccation. In contrast to the nasal horns of centrosaurine horned dinosaurs, which were part of the nasal bones, the nasal horns of *Triceratops* were separate elements from the nasal bones.

Triceratops horridus MARSH, 1889

horridus (HORE-ih-dus) = "dreadful"
Type Specimen: YPM 1820
Referred Specimen: MOR 1110 (skull)

Triceratops horridus had long orbital horns and a relatively larger shield, but an overall shorter skull. Some evidence suggests that *T. prorsus* is the female and *T. horridus* is the male, but Cathy Forster points out that we find a disproportionate number of the two taxa, suggesting that they are more likely different species.

Triceratops was about 7.5 meters (25 feet) long and weighed 4 to 5 tons. Baby *Triceratops* skulls show that their horns developed very early. These baby horns can be mistaken for the head spikes of *Stygimoloch*.

Skulls of *Triceratops* are very common in the Hell Creek Formation, but they are generally found

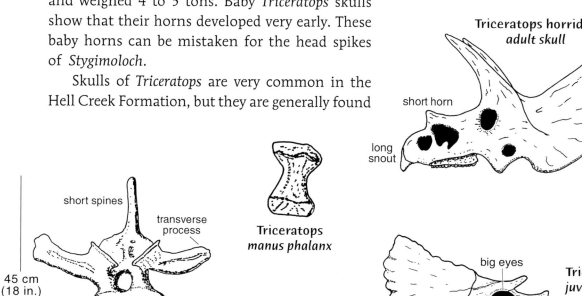

high central ridge

enamel crown

Triceratops teeth

maxillary tooth

4 cm (1.6 in.)

double root

Triceratops humerus

60 cm (24 in.)

massive deltopectoral crest

Triceratops

Triceratops horridus adult skull

no holes in frill

short horn

long snout

short spines

transverse process

45 cm (18 in.)

Triceratops cervical vertebra (anterior view)

Triceratops manus phalanx

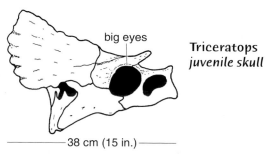

big eyes

Triceratops juvenile skull

38 cm (15 in.)

isolated. Associated or articulated skeletons are rare, and bonebeds are extremely rare. Collectors have found only one bonebed, and the bones were so scattered that there was an average of only one bone per 2 square meters (about 2 square yards). Most bonebeds sampled from Jurassic and Cretaceous sediments contain from five to forty bones per square meter. *Triceratops* teeth are common in microsites.

Torosaurus latus

Torosaurus latus MARSH, 1891
Torosaurus (TORE-o-sore-us) = "piercing reptile";
latus (LAT-us) = "broad, wide"
Type Specimen: YPM 1830
Referred Specimen: MOR 981 (skull)

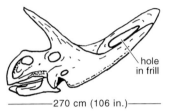

—270 cm (106 in.)—
Torosaurus *skull*

hole in frill

Torosaurus was probably about 6 to 8 meters (20 to 26 feet) long and may have weighed from 4 to 5 tons. The largest skull of any known dinosaur belongs to *Torosaurus*. Some skulls measure nearly 3 meters (10 feet) long.

Torosaurus resembles *Triceratops* except that the neck shield of *Torosaurus* has openings or holes. Some chunks of frill are identifiable because they thin to a feather edge around the two frill openings. The teeth of *Torosaurus*, however, are indistinguishable from those of *Triceratops*.

Immense ceratopsian bones from the Hell Creek Formation are likely those of *Torosaurus*, although there has not yet been a comprehensive study of this dinosaur.

Hadrosauridae

Edmontosaurus annectens

Edmontosaurus annectens MARSH, 1892
(originally called *Claosaurus* and later
changed to *Anatosaurus*, then *Edmontosaurus*)
Edmontosaurus (ed-MONT-o-SORE-us) = reference to the Edmonton Formation; *annectens* (a-NECK-tens) = "joining"
Type Specimen: USNM 2414
Referred Specimen: MOR 003 (skull)

Edmontosaurus is the most common hadrosaur from the Hell Creek Formation of Montana. Articulated and associated skeletons are relatively common, and some specimens are preserved with impressions of skin. The skull can be as much as 1.25 meters (4 feet) long. The animals may have exceeded 12 meters (39 feet) long, possibly weighing as much as 5 tons.

Skeleton of **Edmontosaurus** during excavation from the Hell Creek Formation of south-central Montana (Pvt). Skeleton is 8 meters (26 feet) long.

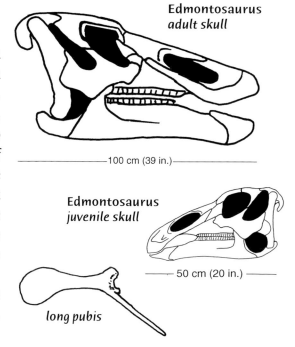

Edmontosaurus adult skull

—————— 100 cm (39 in.) ——————

Edmontosaurus juvenile skull

————— 50 cm (20 in.) —————

long pubis

Edmontosaurus had a much deeper skull than *Anatotitan*, although we don't know how much cranial elongation or depth may reflect old age. The teeth of *Edmontosaurus* are diamond-shaped, and the borders of the enamel have almost no evidence of the crenulations that the teeth of many other hadrosaurs have. Baby and juvenile edmontosaurs have postcranial elements that look identical to the bones of adults. The skulls of the juveniles are much more foreshortened, and the orbits are proportionally much larger than those of adults.

Edmontosaurus remains are often present in large bonebeds in North Dakota, South Dakota,

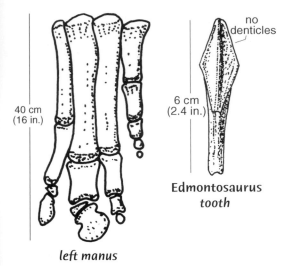

no denticles

40 cm
(16 in.)

6 cm
(2.4 in.)

**Edmontosaurus
tooth**

left manus

and Wyoming. But in Montana, the specimens are usually isolated, associated, or articulated skeletons. No bonebeds have been reported in Montana. Teeth thought to be *Edmontosaurus* are abundant in most microsites.

Anatotitan copei LULL ET WRIGHT, 1942
(originally called *Anatosaurus*, and later changed to *Anatotitan*)
Anatotitan (a-NAT-o-TITE-an) = "giant duck";
copei (COPE-eye) = in honor of E. D. Cope
Type Specimen: AMNH 5730

Anatotitan was about 10 meters (30 feet) long and weighed about 3.5 tons. It may have been semiaquatic, living in lowland swamps.

The skull of *Anatotitan* is extremely long and low. Some skulls that have been referred to *Edmontosaurus* have proportions similar to those of *Anatotitan*, which may suggest that both these taxa are variations within the same genus. The teeth have a slightly elongated diamond shape but are shorter than those of *Edmontosaurus*.

Remains of *Anatotitan* are rare in Montana, found mainly in the southeastern end of the state. *Anatotitan* is common in Wyoming, North Dakota, and South Dakota.

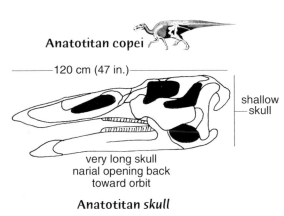

Anatotitan copei

120 cm (47 in.)

shallow skull

very long skull
narial opening back
toward orbit

Anatotitan *skull*

Ankylosauridae

Ankylosaurus magniventris
BROWN, 1908
Ankylosaurus (an-KIE-low-sore-us) = "curved reptile"; *magniventris* (MAG-nih-VEN-tris) = "great belly"
Type Specimen: AMNH 5895
Referred Specimen: MOR 342
(armor plate)

Ankylosaurus was about 5 meters (16 feet) long and about 2 meters (6.5 feet) wide, and it probably weighed as much as 5 tons. This plant eater most likely fed on ferns and other short vegetation.

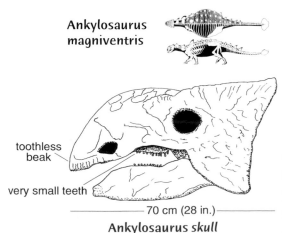

**Ankylosaurus
magniventris**

toothless
beak

very small teeth

70 cm (28 in.)

Ankylosaurus *skull*

Articulated and associated skeletons of *Ankylosaurus*, like those of other armored dinosaurs, are extremely rare, but their isolated teeth and armor plates are fairly common in some strata of the Hell Creek Formation. Among the most identifiable parts of ankylosaurs are the tail vertebrae that connect to the tail club. These vertebrae are fused together to form a massive, rigid rod. The tail club is an enormous, gnarly mass of bone that can approach the size of an automobile tire. If you find one, your biggest problem will be trying to pack it out.

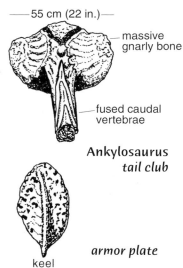

Ankylosaurus tail club

armor plate

The ribs of ankylosaurs are also fused to the vertebrae, and the pelvis is completely fused together. The size of the animal and the extensive fusion of the skeleton make excavation of an *Ankylosaurus* a massive undertaking, particularly if the skeleton is articulated.

Collectors occasionally find *Ankylosaurus* teeth and small armor plates in microsites.

Saurischian Dinosaurs

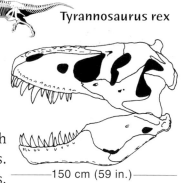

Tyrannosaurus rex

Tyrannosaurus rex OSBORN, 1905
Tyrannosaurus (tie-RAN-o-sore-us) = "tyrant reptile"; *rex* (rex) = "king"
Type Specimen: CM 9380 (originally AMNH 973)
Referred Specimen: MOR 555 (articulated skeleton)
Discovered by: Barnum Brown, 1903
(Includes *Nanotyrannus lancensis*)

Some people consider *Tyrannosaurus* a rare dinosaur, even though collectors have now discovered more than twenty-five skeletons. Seven partial skeletons are reposited at the Museum of the Rockies, as well as numerous individual bones and teeth.

Tyrannosaurus skull

Tyrannosaurus rex reached nearly 14 meters (40 feet) long and weighed around 5,400 kilograms (12,000 pounds). *T. rex* was a carnivorous dinosaur that probably ate mainly *Triceratops* and *Edmontosaurus*. I think *T. rex* was primarily a scavenger, feeding on animals that had died in catastrophic events or stealing carcasses from other predators.

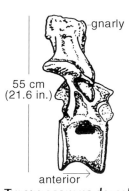

T. rex bones are relatively easy to identify because they are very large and very dense. An average *T. rex* femur measures more than 1.5 meters (5 feet) long and 25 centimeters (10 inches) in diameter. The neural spines of the vertebrae are very intricate and thin, and they are seldom preserved. However, we can readily identify

Tyrannosaurus dorsal vertebra in lateral view

Tyrannosaurus *skull from the Hell Creek Formation, McCone County, Montana (ACE). Skull is 1.5 meters (5 feet) long.*

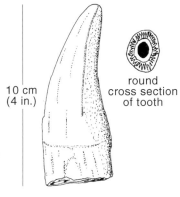

10 cm
(4 in.)

round
cross section
of tooth

Tyrannosaurus *tooth*

the centra even from fragments because the insides look like a honeycomb.

An unbroken *T. rex* tooth can be up to 28 centimeters (11 inches) long and as much as 5 centimeters (2 inches) in diameter. Serrations cover the front and back of each tooth except for the front teeth, which only have serrations on the back sides. The denticles are very small for the size of the tooth.

Nanotyrannus, originally described as a small tyrannosaurid, has more recently been shown to represent a juvenile *Tyrannosaurus*. Differences in the shape of the skulls illustrate the changes that dinosaur skulls undergo during growth.

Tyrannosaurus rex specimens are typically found articulated and usually isolated. *T. rex* teeth are uncommon in microsites. Some collectors report *Albertosaurus* remains from the Hell Creek Formation of Montana, although I think these specimens represent another growth stage of *Tyrannosaurus* rather than a different taxon.

Ornithomimus velox

Ornithomimus velox MARSH, 1890

Ornithomimus (or-nith-o-MIME-us) = "bird mimic";
velox (VEE-lox) = "swift"
Type Specimen: YPM 548
Referred Specimen: MOR 1105 (associated partial skeleton)

Ornithomimosaurs, the group to which *Ornithomimus* belongs, are an intriguing group of dinosaurs because of their characteristic combination of features. They were relatively large yet very gracile, or lightly built. They were built for running, yet possessed no obvious means for catching prey, because their claws are not particularly recurved. They had very large eyes and large brains, possibly as large as those of living ratites such as ostriches and emus. Some paleontologists thought these toothless dinosaurs may have eaten eggs or small prey, but more recent research suggests that ornithomimosaurs were plant eaters.

Most of the skeletal elements of *Ornithomimus* are extremely rare in the Hell Creek Formation, but their foot bones are relatively common. This might be because their vertebrae and leg bones were hollow, whereas their foot bones, although hollow, were more substantial.

The easiest bones to identify from *Ornithomimus* are its finger and toe bones and claws. The toe bones are hollow inside, elongate, and somewhat flattened on their bottom surfaces. The claws from both the hands and feet are fairly straight, long, and hollow. All dinosaur claws carry grooves, sometimes called blood grooves, on their sides, but in *Ornithomimus* these grooves are deep and very distinctive. The surface of ornithomimid bone is very dense and shiny and does not appear porous.

Associated skeletons of *Ornithomimus* are moderately common, but they are also poorly preserved. Isolated elements, particularly phalanges, are present in microsites.

Elmisaurid (genus and species unknown)
Referred Specimen: MOR 752 (articulated foot)

Elmisaurids, or oviraptorids, are extremely rare in Montana, but the one specimen of a foot from eastern Montana indicates they were present. Their skeletal elements look similar to those of other small theropod dinosaurs, or even some birds. Elmisaurid bones are extremely lightly built.

Richardoestesia sp. (species unidentified)
Richardoestesia (ri-card-o-EES-a-zha) = in honor of Richard Estes

Teeth of *Richardoestesia* are common in microsites. A fairly straight shape and minute serrations restricted primarily to their posterior edges identify the teeth. The teeth carry are from 7 to 10 denticles per millimeter (178 to 254 denticles per inch). The teeth measure only about 7 or 8 millimeters (about 0.25 inch) long and are shaped like an isosceles triangle.

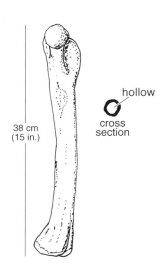

38 cm (15 in.)

hollow

cross section

Ornithomimus *left femur*

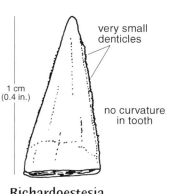

very small denticles

1 cm (0.4 in.)

no curvature in tooth

Richardoestesia *tooth*

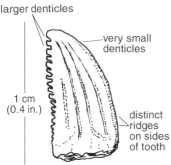

larger denticles

very small
denticles

1 cm
(0.4 in.)

distinct
ridges
on sides
of tooth

Paronychodon *tooth*

Theropod Teeth of Unknown Origin

Most microsites from the Hell Creek Formation of Montana yield a number of interesting but unidentifiable small theropod teeth.

"*Paronychodon*" is a name Cope coined in 1876 in reference to some theropod teeth from the Judith River Formation. They have a flattened lateral surface. Collectors have since found similar teeth in various formations in Canada and Montana, including the Hell Creek Formation. O. C. Marsh thought these teeth sat at the anterior ends of dentaries in such a way that the two teeth fitted against one another. Phil Currie and others infer that the teeth may be the result of abnormal growth. Teeth with similar morphology have been found in the jaws of other theropod dinosaurs, including *Saurornitholestes* and *Troodon*.

Various other unidentified theropod teeth are known from the Hell Creek Formation. They will probably remain unidentified until someone finds them in a jaw associated with some cranial material.

Aves

Bird remains are uncommon in the Hell Creek Formation, probably because many of their bones are likely indistinguishable from those of small or juvenile nonavian theropods. The tarsometatarsus is one of the few bones that we can readily identify. The most common remains are tarsometatarsi referable to the diving bird *Hesperornis*.

Common Nondinosaurian Vertebrate Fossils from the Hell Creek Formation

The remains of sharks and bony fishes are relatively abundant, mainly as isolated scales and skull parts, in the stream and river deposits of the Hell Creek Formation of Montana.

There are a variety of elasmobranch, or shark, taxa known from the Hell Creek Formation, but one taxon, *Myledaphus bipartitus*, is particularly abundant. Its teeth and vertebrae easily identify it.

The remains of bony fishes, or Osteichthyes, are common in most fluvial systems of the Hell Creek Formation, and are primarily represented by scales, skull parts, and vertebrae. The most abundant are the scales of garfish (*Lepisosteus occidentalis*), although their skull elements and vertebrae are also common. Another relatively common fish is the sturgeon *Acipenser albertensis*, represented by osteoscutes and fin spines.

Amphibian remains are relatively rare in the Hell Creek Formation, although collectors occasionally find some vertebrae that we can readily identify as *Scapherpeton tectum*. These vertebrae have a very distinct ventral keel.

Numerous taxa of turtles are present in the Hell Creek Formation. The most common turtles are the trionychids, or soft-shelled turtles. Their shells can measure as much as 50 centimeters (20 inches) long. *Trionyx* is the most common of this group. Another common variety is the baenids, aquatic turtles that had massive jaws, probably for eating mollusks. Also present are dermatemydids, of which *Adocus* is a reasonably easy aquatic taxon to find and identify. *Basilemys*, a terrestrial tortoiselike taxon, is less common. The cryptodiran turtle *Compsemys* might be the easiest turtle to identify, because small, raised tubercles cover the surface of its carapace.

Champsosaurus rib in anterior view (actual size)

Some of the more common nondinosaurian reptiles from the Hell Creek Formation are the champsosaurs, aquatic gavial-like animals (long-snouted crocodiles) about 1 meter (3 feet) long. Vertebrae of *Champsosaurus* are about the size of the end of a person's thumb and are distinctive. The neural arch is seldom fused to the centrum, and the dorsal, or top, surface of the centrum has a distinctive hourglass pattern. The ribs of champsosaurs are also very easy to identify.

Crocodiles from the Hell Creek Formation look much like modern crocodilians except that the fossil specimens are smaller. The most common crocodilian, *Leidyosuchus,* is only slightly more than 1 meter (3 feet) long. Crocodilian skull bones stand out because they have highly pitted and rough surfaces. You can easily identify crocodilian vertebrae because they are either strongly opisthocelous or procelous and about the size of the end of a person's thumb.

Collectors have found various mammal taxa in the Hell Creek Formation. The most abundant and identifiable are the multituberculates—small, rodentlike forms with unique, blade-shaped lower cheek teeth. Most mammal remains consist of jaws with or without teeth and isolated teeth. They are most common in microsites.

Pseudofossils
Dinosaur Bone and Egg Look-Alikes

PEOPLE BRING A LOT OF DIFFERENT THINGS to museums for identification, and many people are pretty sure what they have even before coming to the museum. Sometimes people are just looking for someone to confirm their suspicions. A lot of the dinosaur "eggs" and "bones" people find turn out to be rock look-alikes and not fossils at all. And rock look-alikes fool not just the general public, but trained paleontologists. I once found what I thought was a turtle shell and brought it back to the museum for study. I soon discovered that I had actually found a chunk of Native American pottery. We call these look-alikes pseudofossils. The most common are spherical or elongated sandstone and limestone concretions, geodes, and septarian nodules. At the Museum of the Rockies, we save and curate some of the best of these pseudofossils so we can show people what are not fossil eggs and bones.

Pseudofossils are interesting rocks, and like fossils, are worthy of collection and display. But they are not fossils. You can distinguish between these pseudofossils and the real thing by looking closely for the structures of bones and eggs. Bones, whether permineralized or replaced, retain their inner structures, such as the spongy marrow spaces. Fossilized dinosaur eggs also retain their distinctive structure, with a dense outer layer of calcium where the shell was and a less dense and unstructured interior.

Concretions
Concretions are rocks that can fool even an eye trained to distinguish dinosaur fossils. Concretions exist in many of the same formations that contain dinosaur fossils. They form in many kinds of sediment, including shale, sandstone, and limestone.

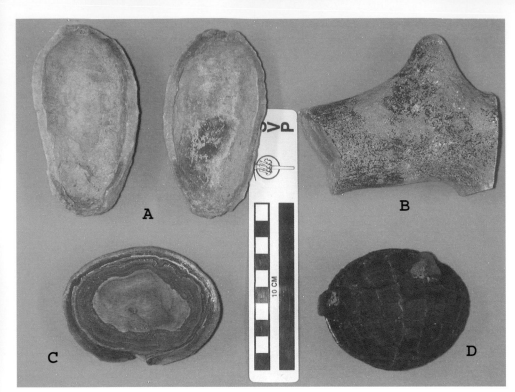

Various concretions and nodules that look like fossil eggs or bones. A, a hollow sandstone concretion split open; B, an elongated, bonelike sandstone concretion; C, a banded-ironstone concretion; D, an ironstone septarian nodule.

Concretions are usually brown or black because they predominantly contain iron minerals. Concretions come in three shapes: spheres, ellipses, or cylinders. They can range from the size of a pea to many feet in diameter. Some concretions develop around the outside of real fossils or other objects, but most are simply odd-shaped rocks. Most of the "fossil eggs" that people bring to the museum for identification turn out to be concretions.

The best way to identify a concretion is to look inside. Most concretions are layered, with concentric rings on the inside similar to the rings in a tree. Cylindrical concretions might at first resemble bones, so look closely at the structure. On the inside of a dinosaur bone, you can see the marrow, or more accurately, the spongy bone where the marrow resided. However, on the inside of a concretion, you will see neither structure nor concentric rings. Spherical and ellipsoidal concretions may look like dinosaur eggs, but they lack the texture or composition of eggs. Dinosaur eggs have a single layer of calcitic eggshell that typically looks dense and is a different color from the rock that fills the inside. In contrast, spherical and ellipsoidal concretions commonly look sandy on the outside and have characteristic concentric rings on the inside.

An ironstone septarian nodule exhibiting a checkerboard surface. The nodule is about 8.5 centimeters (3 inches) long.

If you find what you think might be a fossil egg and not a concretion, take it to a museum for positive identification. It may very well be an important scientific discovery. But if your find turns out to be a concretion, don't feel bad—concretions fool even the best of us. I put concretions in lab exams for my students, and most of them identify the concretions as eggs or bones because they forget to examine the insides.

The Hell Creek Formation commonly contains small ironstone concretions about the size of a person's fist. These concretions look like small cannon balls. Concretions are rarely found in the Cloverly and Judith River Formations. The Hell Creek Formation holds large concretions, some as large as a pickup truck, in many areas.

Septarian Nodules

Septarian nodules are unusual rocks. They look like concretions that were cracked and then filled with a crystalline mineral. When weathered, these rocks can closely resemble vertebrae or other parts of a skeleton. Like concretions, nodules lack the internal structures that identify bones.

Collectors occasionally find septarian nodules in the Hell Creek Formation, but they are most abundant in marine sediments.

Geodes

Geodes are oddly shaped rocks that may resemble bones and teeth. Some of these rocks consist of layers of different hardnesses and, therefore, weather oddly. Others consist of sediments of different colors. In either case, you can distinguish geodes from bones or teeth by their internal structure.

Like some concretions, geodes can be spherical, but geodes are usually hollow, and their inside surface may contain crystals. The insides of bones have spongy marrow structures, and dinosaur teeth consist of dentine and enamel, like teeth of living animals. Whereas concretions develop from the center outward, geodes form from the outside inward. Geodes are rare in most of the terrestrial sediments of Montana but are present in some marine rocks.

Odd Rocks

Sometimes ordinary rocks weather in the shapes of fossils, but on close examination you can easily tell the difference. Some limestone rocks that contain mats of mineral-replaced algae weather unevenly. The weathering patterns turn chunks of the limestone into look-alikes for vertebrae or other bones. The different minerals that give some rocks their variegated color combine with weathering patterns to make the rocks look like teeth or bones.

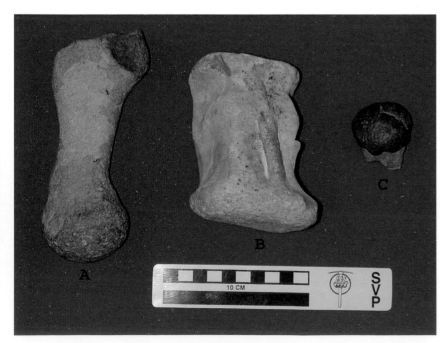

Three odd rocks that look like fossils. A, a sandstone rock that has weathered to look like a limb bone; B, a limestone rock that contains the fossilized remains of algae that resembles a vertebra; and C, a two-colored rock that resembles a tooth.

Collection, Preservation, and Curation of Vertebrate Fossils

THE MOST IMPORTANT THINGS ABOUT ANY FOSSIL are its care and the information about its geographical and geological location. If you discover what you think is the skeleton of a dinosaur or any other extinct animal, the best thing you can do is to call a trained paleontologist who will come and excavate the specimen, and then take it back to a laboratory for preparation and curation. You must follow this procedure if you found the fossil on federal or public land. If, however, you have discovered a fossil on private land, and you decide you would like to collect, prepare, and care for the fossil, there are a few procedures that you should follow to ensure that the specimen is of use to science.

Collection and Preservation

If you decide to try to collect dinosaur remains yourself rather than calling in someone experienced in paleontological collection, consider how fragile the fossils may be. Fossils lying on the surface of the ground are usually already fragmentary, and you generally can pick them up without further damaging them. However, if the fossils are partly in the rock you must take extreme care in removing them.

Although most fossils are petrified or permineralized, some dinosaur remains are only partially fossilized. So before picking up a dinosaur bone, treat it with some kind of hardener or glue. The preparators at the Museum of the Rockies use a glue called Vinac or another substance called Paleo-Bond. As a temporary substitute, you can also use watered-down white glue, such as Elmer's. Most importantly, get the glue thin enough to soak into the bone, as any glue you use MUST soak into the bone to do any good. Then wait until the glue has hardened—an hour or two—before you attempt to lift the specimen from the ground. Patience is essential for paleontologists. If it's a large bone, you might have to dig around it and make a splint for it.

Before removing a fossil from the ground, you will need to jacket it in plaster. First, make sure the bone has been hardened with glue. Then pedestal the bone by digging around and slightly under it. After the bone is hard from gluing and it is on its pedestal, place wet paper towels over the entire surface to keep the wet plaster, which you'll soon apply, from touching the bone. Then dip strips of burlap or another open-knit cloth in wet plaster of paris and tightly fit them around the bone, over the paper towels. You will need to wrap large pieces in more than one layer of plaster strips, and it's a good idea to stiffen the package with boards. The plaster jacket must be strong enough that it doesn't bend or break when you pick it up after it is dry. But you also don't want to cover the specimen with so much plaster that it becomes too difficult to get off later without damaging the fossil. Determining the right amount takes experience.

When the plaster sets, it will encase the bone much like a cast on a broken arm or leg that a doctor has immobilized. You can then turn over the plaster package and cover its bottom side with cloth strips and plaster. After this side has set, the package is ready for the trip to the lab for final cleaning and curation.

You can best collect material from microsites by screen-washing the sediments. Screen boxes are about 8 to 10 inches deep, and a screen mesh covers their bottoms. For best results, particularly for catching small mammal teeth, use a screen size smaller than that found on an ordinary screen door. Geology supply houses commonly carry small-mesh brass screen that works well for this.

Dump sediment into the boxes and let it sit overnight in water at the edge of a stream or lake to allow the sediments to break apart. Then gently swish water back and forth to drain the mud, leaving the fossils and rocks. Dump this matrix out on a dry surface and allow it to dry before picking out the fossils.

Once the fossil is home or in a lab, you can prepare your fossil specimen. Small tools work best for this. Use an old toothbrush and a small paintbrush to remove some dirt and loose rock. For harder sediment, you might have to use a needle, a small

Photographic record showing the steps to jacket a fossil in the field.

A, excavating the specimen;

B, initial wrapping of the specimen in wet paper towels;

C, specimen is completely wrapped in paper towels;

D, adding plaster-dipped burlap strips to the towel-covered surface;

E, completed plaster jacket drying and setting.

Bob Harmon and Mary Schweitzer clean rock off the vertebrae of Tyrannosaurus rex. —Museum of the Rockies photo

drill or Dremel tool, or an old dental pick (ask your dentist if he or she has some extras). When using metal tools, always take care to not scratch or gouge the surface of the bone—or yourself. If you do gouge the bone, make a note of it so others won't mistake the gouge for a bite mark or some other natural scar. At the Museum of the Rockies, the preparators use small air scribes, air abrasive units, and in some cases, even various acids to slowly eat away the rock.

For preparing small specimens, always use some kind of magnification device, such as a microscope. That will help you minimize the chance of gouging the fossil and allow you to do a good job of removing the rock from detailed areas.

When you have cleaned your fossil, write the specimen number on it with India or other permanent ink. If the bone is dark, paint a small white stripe on the bone, and then write the number on the stripe. For your collection to be of any scientific use, specimen numbers are essential.

Curation

Curation is the process of identifying and caring for the specimen once it has been prepared for study or display. As a curator, it is my job to identify the specimens once the preparators have finished their cleaning and gluing. Once I have identified a specimen, I log it into our collection catalogue, giving the specimen its own museum number and recording all the field

Name ___DASPLETOSAURUS___	BfN
Museum No. MOR ___590___	
Field No. ___8-11-89-2___	Date ___1989___
Geology ___Two Medicine Formation (upper ¼)___	
Locality ___TM-068 Glacier County, MT___	
Description ___Skull and Leg___	
Collector ___V. Clouse & B. Harmon___ Preparator ___B. Harmon___	
Acc No: ___V.89.1___ Location ___C.1.5___	
Museum of the Rockies Montana State University	

An example of a good specimen label

and collection information that came with it. I also assign a locality number, and with this document is all the geological and geographical data that was acquired with the specimen. One of the most important curatorial steps is placing the museum number on the specimen. We do this by painting on a small white stripe on the specimen and then writing the museum number on the stripe using India ink, which is stable and durable. When the specimen is numbered and a paper label has been made with all of the specimen's pertinent information, the specimen and label are placed in a shallow, foam-padded box and put away in a dust-proof case. The location of the specimen is recorded in the museum collection catalogue so anyone who wants to study the specimen can locate it.

Every specimen you have in your collection will have a unique *specimen number,* and your collection should start with Specimen Number 1. At the Museum of the Rockies, we started with 001. The *specimen name* is the name of the animal. Typically your first guess will be wrong—mine often are—so write the name in pencil. *Description* is the name of the different part or parts you found of the animal. The *locality* is the name of the landowners, with their phone number, and the exact location of your find. Also include references to the field notes the collectors took, copies of which should be kept in a permanent repository. This information must be good enough to guide someone else to your site. The *geology* is whatever you can say about the rocks in which you found your specimen. Include color and texture if you can. Any information is good. The *collector* should include everyone who was present at the time the specimen was collected. The *date* is the date the specimen was found, regardless of when it might have been collected. The *field number* is the number you assign to the specimen in the field. On the accompanying label, the field number consists of the date and the number of the specimen for the day. So the field number tells us that the *Daspletosaurus* skull and leg bone were collected on August 11, 1989, and were the second specimens found and recorded that day. The field number is particularly important if you are collecting a lot of specimens or if your specimen has been put in a bag or plaster jacket—it is easy to forget what's in the jacket or paper bag. You may wish to record *field notes,* which include the kind of glue you used, how you wrapped the specimen, and any other information you think might be useful. *Preparation* includes what you did to the specimen after you brought it back to your "lab." Make a note of the glue you used to put broken pieces back together and what kind of glue you used to harden the bone. Also note the name of the preparator and the storage location of the specimen.

If you have a drawer in which to store your specimens, be sure it has a soft surface. At the Museum of the Rockies, we lay thin pads of foam in drawers, in boxes, and on shelves for the fossils to rest on, and we put small specimens in small, padded plastic boxes.

As the curator of paleontology at the Museum of the Rockies, I direct the research, collection, and care of all the specimens in the Paleontology Department. We store our paleontology collection on a database that you can access through the World Wide Web at **www.museum.montana.edu.** Please feel free to browse through the collection. The database is searchable, so if you are interested in a particular kind of dinosaur or other fossil, you can see if we have it. As we photograph this collection, we place the pictures on our web site so to help other collectors identify fossils. If you have any questions about collecting, curating, or dinosaur research, contact the Paleontology Department at the Museum of the Rockies, Montana State University, Bozeman, Montana.

Museums and Dinosaur
Dig Sites in Montana

MOST MUSEUMS ARE BUILT as places to display things from our past, both historic and prehistoric, for enjoyment and education. But there are two kinds of museums: the display museum and the repository museum. Repository museums store historic and prehistoric items for safekeeping, research, and future display. The Museum of the Rockies is both a display museum and a repository museum. Some of our museum items are on display where people can see them, but many, many more items are in the repository collection. The Museum of the Rockies has hundreds of thousands of items, and of those, about 15,000 are fossils. It would take a display museum twenty times the size of our present facility to display all our reposited collections, so we keep most of them safe in climate-controlled collection rooms in the basement. The basement also has research rooms where my colleagues from other museums, my students, and I can study the fossils without bothering people visiting the displays. Historians, scientists, and even interested amateurs can make appointments to see and study these collections. The best specimens are either on display or being studied and will go on display when the studies are completed.

When amateur and professional paleontologists or collectors from the general public find fossils on land owned by our federal government or on Montana state land, the specimens must be placed in a repository. The two repositories for fossils in Montana are the Museum of the Rockies at Montana State University in Bozeman and the Department of Geology at the University of Montana in Missoula. Neither of these repositories owns the fossils—all fossils found on public lands belong to the people of the United States. Our museums just take care of the fossils.

As a state repository, the Museum of the Rockies has an agreement with both the Montana Department of Natural Resources and Conservation and the Montana Department of Fish, Wildlife and Parks to commit the funds required to store and safeguard specimens found on state property—and to do so in perpetuity. As a federal repository, the museum has agreements with all the agencies of the federal government that have lands in Montana to store and safeguard their specimens in perpetuity. What this means is that the Museum of the Rockies, or any other museum that makes this kind of commitment, is responsible to care for these specimens with or without monetary help from either the state or the federal agencies.

Another Montana repository is the Montana Historical Society Museum in Helena. As our state museum, the Historical Society cares primarily for many of our historical and archaeological artifacts. Some other museums in Montana care for and display specialized collections. For instance, the Charlie Russell Museum in Great Falls cares for Russell's house and many of his paintings.

In Montana we are very fortunate in also having a great many local museums—one in almost every county. Some of these local museums display dinosaurs. The Carter County Museum in Ekalaka, for example, has displayed dinosaur skeletons for many years, thanks to its curator, Marshall Lambert.

Montana Museums that Display Dinosaur Fossils

Museum of the Rockies, Montana State University, Bozeman
Regional museum showing numerous dinosaur specimens from Montana and northern Wyoming. Houses one of the largest dinosaur collections in the United States.

Department of Geology, University of Montana, Missoula
Small displays of fossils from Montana.

Carter County Museum, Ekalaka
Mounted skeleton of an *Anatotitan* and skulls of *Triceratops* and other dinosaurs, all from Carter County.

Fort Peck Interpretive Center, Fort Peck
Tyrannosaurus rex skeleton being prepared on display, plus other local dinosaur fossils.

Teton Trail Museum, Choteau
Maiasaura and *Saurornitholestes* skeletons and other dinosaur fossils from the Two Medicine Formation.

Hill County Museum, Havre
Portions of *Albertosaurus* skeleton and numerous other dinosaur remains, including eggs from Hill County.

Phillips County Museum, Malta
Cast skeleton of *Albertosaurus,* skeleton of *Brachylophosaurus,* and other dinosaur remains from Phillips County.

Garfield County Museum, Jordan
Cast skeleton of *Triceratops* and skull of *T. rex*, and other dinosaur fossils from the area.

Makoshika State Park, Glendive
Triceratops skull and several other fossils found in the park.

Blaine County Museum, Chinook
Cast of *Albertosaurus* skull and many dinosaur remains from Blaine County and local area.

Where You Can See Dinosaur Fossils Being Excavated

Nature Conservancy's Egg Mountain Dinosaur Preserve, Choteau

Dinosaur Digs—Field Schools and Pay-to-Dig Sites

Montana State University, Havre
Assisting the research efforts of Vicki Clouse; research changes from year to year.

Teton Trail Museum, Choteau
Digs near Egg Mountain and Choteau.

Time Scale, Inc, Bynum
Digs near Choteau; pay-to-dig field school that assists paleontological research.

Appendix I

Some State and Federal Agencies that Manage Land in Montana

Federal Agencies

Bureau of Land Management (BLM)

Billings Field Office
P.O. Box 36800
(5001 Southgate Drive)
Billings, MT 59107-6800

Lewistown Field Office
P.O. Box 1160
(Airport Road)
Lewistown, MT 59457-1160

Havre Field Station
1704 Second Street West
Drawer 911
Havre, MT 59501-0911

Bureau of Reclamation (BR)

Montana Area Office—Great Plains Region
P.O. Box 30137 (2900 Fourth Avenue North)
Billings, MT 59107

United States Fish and Wildlife Service (USFWS)
P.O. Box 110
Lewistown, MT 59457

United States Forest Service (USFS)
Region 1
P.O. Box 7669
Federal Building
Missoula, MT 59807

State Agencies

Montana Department of Natural Resources and Conservation (DNRC)

Lands Division
P.O. Box 201601
(1625 Eleventh Avenue)
Helena, MT 59620-1601

Montana Department of Fish, Wildlife and Parks (MDFWP)

P.O. Box 200701
(1420 East Sixth Street)
Helena, MT 59620-0701

Appendix II
Skeletal Details

General Skeletal Information

All animals with a backbone, or vertebrates, have the same basic skeleton with a head at one end, a tail or tailbone at the opposite end, and four or fewer appendages in between. Vertebrate animals are also bilaterally symmetrical; that is, the right side of the skull and skeleton is a mirror image of the left side.

We use special words for anatomical directions and orientation when describing skulls, skeletons, or individual bones. These terms have been in use for centuries. The head and tail ends of an animal or a single bone are the anterior (front) and posterior (back) directions, respectively. Some books refer to the anterior, or head, direction as cranial and the posterior, or tail, direction as caudal, but these are uncommon directional terms. Words that refer to the directions toward the back or toward the belly are dorsal (up) and ventral (down), respectively. You can combine many of these terms to give more precise directions. For example, a bone located in both an anterior and ventral direction is in

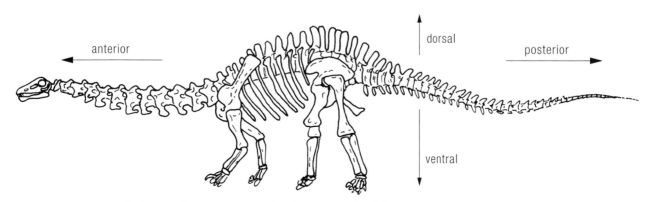

Skeleton of Apatosaurus showing anatomical directions

155

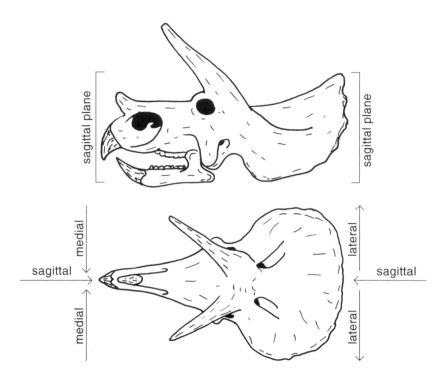

Skull of Triceratops in lateral and dorsal views, showing saggital, medial, and lateral anatomical directions

the antero-ventral direction. Postero-dorsal is in the posterior and dorsal directions.

Transverse (side to side) and sagittal (in a vertical plane) directions are more difficult to imagine. Place your hand in front of your face with your palm parallel to the floor. Move your hand up against your face, and imagine a plane parallel to your hand that goes through your head. That plane is the transverse plane or transverse direction of your head. Now turn your hand so your fingertips point toward the ceiling. Move your hand to the center of your face and imagine a plane parallel to your hand that goes through your head. That is the sagittal plane of your head. Moving toward the body midline, or sagittal plane, is the medial direction, and moving outward from the midline is the lateral direction. The heart is medial, or inward, to the arms, and the arms are lateral, or outward, to the heart.

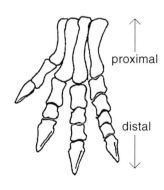

Foot of Tenontosaurus showing proximal and distal anatomical directions

Two other directional terms that we commonly use when discussing bones are proximal and distal. The proximal direction is toward the body, and the distal direction is away from the body. If you look at your thigh bone, or femur, the proximal end attaches at your hip—it is closest to the center of your body. The distal end forms the knee—it is the point of the thigh farthest from the center of your body.

At the anterior end of the skeleton is the skull, and posterior to, or behind, the skull are two or fewer sets of limbs. If limbs are present, they typically consist, of three segments: the proximal unit, composed of a single bone; the middle unit, composed of two bones; and the distal unit, which can have various elements. On a typical hind limb the proximal unit includes the thigh bone, or femur. The middle unit includes the shinbones, or tibia and fibula. The distal unit is the hind foot and toes, collectively called the pes.

On all vertebrates except some basal tetrapods, the elbow always bends backward and the knee always bends forward and commonly outward. If you look at a bird, its legs, which are hind legs, appear to bend the wrong way. That's because you are looking at the ankle joint rather than the knee. As we know from looking at a chicken, muscle completely binds the thigh and feathers cover it, so you can't see the proximal unit of the leg when you look at a live bird. The leg that you can see on a live chicken consists of the tibia and fibula (drumstick), the ankle, and the foot. The femur is hidden in muscle and feathers. The leg joints of a horse also look like they bend the wrong way. This, too, is because muscles bury the knee and elbow joints. The joints that we can easily see on a horse arc the ankles and wrists.

Skull

The skull is obviously the most important part of a dinosaur skeleton because it is the most characteristic. With a skull, you can determine which species a skeleton belongs to or if it's a new species. The skull of a dinosaur consists of many different bones, each of which you may find isolated or separated from other cranial bones.

Dinosaurs and other reptiles and birds have many different bones in their lower jaws. We mammals have only one, the dentary. On dinosaurs, the dentary is the bone of the lower jaw that contains the teeth, and the maxilla is one of two bones in the upper jaw that may have teeth. The premaxilla of most dinosaurs is toothed, but in some dinosaurs this bone forms a toothless beak or bill instead. The premaxilla forms the ducklike upper bill of the duck-bill dinosaurs, or hadrosaurs. Horned dinosaurs, or ceratopsians, have an extra bone attached to their premaxilla

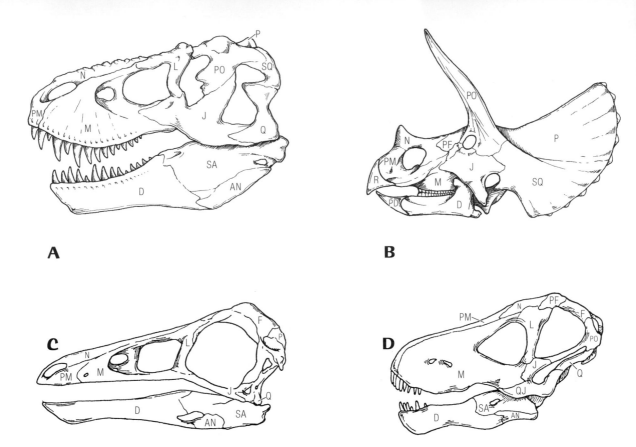

The cranial bones as they appear on four different dinosaur skulls. A, Tyrannosaurus; B, Triceratops; C, Ornithomimus; D, Diplodocus

Abbreviations for skull bones:

AN=angular	*L=lacrimal*	*PM=premaxilla*	*R=rostral*
BO=basioccipital	*M=maxilla*	*PO=postorbital*	*SA=surangular*
D=dentary	*N=nasal*	*PR=prefrontal*	*SQ=squamosal*
F=frontal	*P=parietal*	*Q=quadrate*	
J=jugal	*PD=predentary*	*QJ=quadratojugal*	

called the rostral, which forms the bony part of their birdlike beak. Attached to the beaks or bills of all the plant-eating dinosaurs was a horny material made of the same substance as our fingernails. The scientific name for this horny material attached to the beak is rhamphotheca. Ornithischian dinosaurs have a unique bone called a predentary that connects the two lower jaws together. This bone apparently had a rhamphotheca similar to the one on the premaxilla.

In all dinosaurs, the premaxilla and nasal bone join to form the nose hole, or narial opening. The nasal bone is also important in some horned dinosaurs because part of it forms the nasal horn. In duck-bill dinosaurs, the nasal bones may form elaborate crests, some of which are hollow. The dorsal surface of the nasals of tyrannosaurs are rough and gnarly. Some paleontologists think that rough area might have supported a hornlike protuberance.

Another important bone on the skull is the jugal, which forms the lower boundary of the eye hole, or orbit, and often the anterior and ventral boundaries of an opening called the lateral temporal fenestra. The purpose of this opening is unclear, but it is present in almost all dinosaurs, most reptiles, and all birds. The only dinosaurs that don't have this opening are the armored dinosaurs, and armor apparently secondarily covered their lateral temporal fenestrae. In the orbit of all dinosaurs, a series of thin bones attached to one another to form a circle that surrounded the eyeball. These small bones are called sclerotics. They are very difficult to identify if isolated from a skull.

The postorbital bone attaches to the jugal. It forms the upper posterior part of the orbit and the anterior and dorsal borders of the lateral temporal fenestra. It also forms the anterolateral side of the supratemporal fenestra. The postorbital bone forms the orbital horn in horned dinosaurs. Other bones important in the skulls of horned dinosaurs include the squamosal and parietal bones. These bones form the neck frill of the horned dinosaur's skull. In other dinosaurs, the parietal simply covers the top of the brain and forms the medial side of the supratemporal fenestra. The squamosal forms the posterodorsal corner of the lateral temporal fenestra and the lateral posterior corner of the supratemporal fenestra. The squamosal of all dinosaurs is on the posterodorsal corner of the skull.

The quadrate is a long bone that fits between the squamosal and the articular bone of the lower jaw. It is the largest bone that connects to the lower jaw at the hinge of the jaw. The articular is a very small bone that rests on either the prearticular or surangular bone of the lower jaw.

The dinosaur braincase is typically small, and a number of nerve openings penetrate it. In rare instances, the braincase falls apart to reveal the individual bones, called cranial elements. Skull bones vary greatly from one dinosaur to another, and it is very difficult for even an experienced paleontologist to identify these

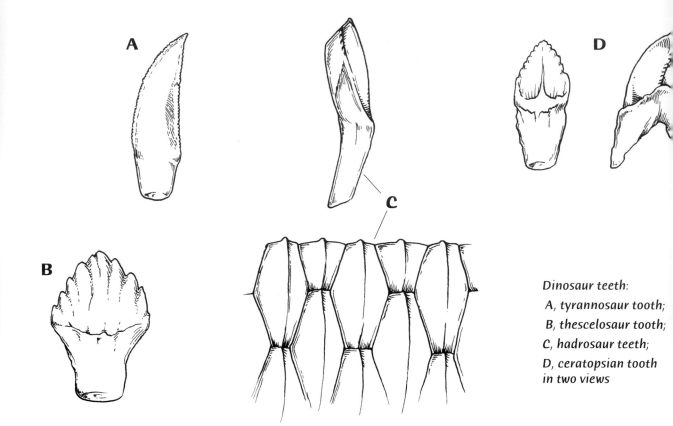

A

B

D

C

Dinosaur teeth:
A, tyrannosaur tooth;
B, thescelosaur tooth;
C, hadrosaur teeth;
D, ceratopsian tooth
in two views

elements when found individually. (For technical drawings of these bones in a duck-bill dinosaur, see the scientific paper by J. R. Horner, 1992, listed in Additional Reading.)

The basioccipital, on the posteroventral side of the skull, has a ball-shaped surface that connects to the first vertebra. The ball is called the occipital condyle. All reptiles, including dinosaurs and birds as well as fishes, have a single occipital condyle, whereas amphibians and mammals have a double condyle. Attached to the sides of the basioccipital are a pair of exoccipitals that form the posterior end of the skull. Between the exoccipitals and forming the dorsal and posterior end of the braincase is the supraoccipital. Anterior to this bone is the parietal. Paired frontals attach to the anterior end of the parietal and form the roof over the anterior end of the brain. On the anterior end of the basioccipital is the basisphenoid. This is the bone that forms most of the floor of the braincase and houses the pituitary gland. Resting on the dorsal surface of the junction of the basioccipital and basisphenoid are the prootic bones. Together with the exoccipitals, the prootic bones house the semicircular canals and ear region. Anterior to the prootics are the laterosphenoids. These form the lateral walls of the anterior braincase and connect to the postorbitals. Anterior to the laterosphenoids are the orbitosphenoids and presphenoids, through which the nerves for the eyes and nose pass.

The teeth of all animals have two basic parts, the crown and the root. The crown is the part that is exposed above the gum, and the root lies buried beneath the gum line. All dinosaur teeth, except those belonging to horned (ceratopsian) dinosaurs, are single rooted. Ceratopsian teeth have two roots.

Dinosaur teeth come in two basic varieties, those used for eating meat and those for eating plants. Meat-eating, or carnivorous, dinosaurs are called theropods. Their teeth are typically sharply pointed and have many small serrations or denticles running down one or two sides of the tooth's crown. We can sometimes use the number of serrations to identify the owner of the tooth. Theropod teeth have an enamel layer over the entire surface of the crown, which makes them very shiny. Theropod teeth come in various sizes, depending on the size of the animal and whether it was a juvenile or adult. Most theropod teeth range from about 0.5 to 12 centimeters (0.2 to 4.7 inches) long. Roots can add even more length to them. The roots of theropod teeth usually have the same cross-sectional shape as the crown.

The teeth of plant eaters, or herbivores, are hugely variable. The basic herbivorous dinosaur tooth has a leaflike crown with large serrations along its anterior and posterior sides. Enamel covers the entire surface. The roots of these teeth are round, regardless of the shape of the crown. These teeth are found in both early ornithischian and early sauropodomorph dinosaurs. Many of the advanced plant-eating dinosaurs, particularly hadrosaurs and ceratopsians, had teeth with enamel on only one side and roots that allowed their teeth to be locked together into densely packed dental batteries.

Some dinosaurs, such as the ostrich dinosaurs, or ornithomimids, and all living birds lack teeth. Some primitive birds had teeth.

Dinosaurs replaced their teeth throughout their lives. Consequently, dinosaur teeth are common in sediments that contain dinosaur fossils. Tyrannosaurs kept their teeth for about a year before replacing them, but some dinosaurs, such as the hadrosaurs,

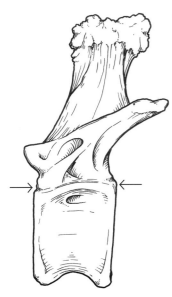

replaced teeth within months. Hadrosaurs had compact dental batteries with hundreds of teeth waiting for other teeth to wear down and fall out.

Vertebrae

The back, or posterior end, of the skull attaches to the vertebrae. The first vertebra in the skeletons of most animals, including amphibians, reptiles, birds, and mammals, is called the atlas. This is the vertebra that allows us to move our head up and down. The second vertebra, called the axis, allows reptiles, birds, and mammals to move their heads from side to side. The atlas lets us nod yes, and the axis lets us say no. Amphibians, such as frogs and salamanders, can't move their heads from side to side because they don't have an axis. These animals have to move their entire bodies to look from side to side.

Vertebrae have two important units, the centrum and the arch. The centrum is the lower, spool-like part that helps strengthen the body. The neural arch above the spool protects the spinal cord and is also an attachment for muscles. The dorsal process, or dorsal end, of the neural arch is called the neural spine. There can be winglike structures on the sides called transverse processes. Four flat, smooth areas, called zygapophyses, on each vertebra are for connection to the adjacent vertebra. The two anterior zygapophyses face upward and inward to receive the two posterior zygapophyses, which face downward and outward. If you are lucky enough to find an articulated series of vertebrae in sediment, you will be able to determine the direction of the animal's head by

The two segments of a tyrannosaur dorsal vertebra: the round centrum (bottom segment) and the neural arch (top segment). The arrows point to the junction of the two segments. Note the openings, called pleurocoels, that penetrate the lateral sides of the centrum and neural arch.

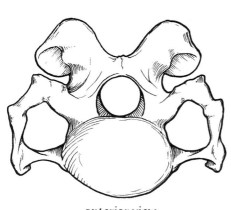

anterior view

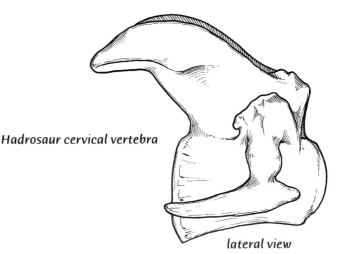

Hadrosaur cervical vertebra

lateral view

examining the position of the zygapophyses. If the anterior end of the vertebrae, with their upward-pointing zygapophyses, points into the hill, there might be a skull in the hill. If the upward-pointing zygapophyses point away from the hill, then the hind end of the animal is probably in the hill.

Dinosaurs have four kinds of vertebrae: the cervicals, dorsals, sacrals, and caudals. The cervical vertebrae make up the neck. All mammals, including humans, whales, and even giraffes, have only seven cervical vertebrae. The number of cervical vertebrae varies between dinosaur species—some species have as few as seven, while others have nearly twenty. Dinosaur cervical vertebrae vary considerably in shape depending on the species, but generally the centrum is oval and the neural arch is wide with large processes, or ends. The centrum is usually opisthocelous, which means it is deeply concave on the posterior end and highly convex on the anterior end. Small cervical ribs attach to the vertebrae of all dinosaurs. In sauropods, the cervical ribs can be very long, and they probably supported the throat and its muscles.

Hadrosaur dorsal vertebra in posterior view

The dorsal vertebrae support connection of the ribs. Dinosaurs can have fewer than ten or as many as fifteen dorsal vertebrae. The dorsal vertebrae of mammals are divided into thoracic and lumbar vertebrae because the anterior and posterior dorsals are characteristically different. Ribs connect to the thoracic vertebrae of mammals. Dinosaur dorsal vertebra centra are amphiplatyan, which means that they are flat at both the anterior and posterior ends. In theropod dinosaurs, openings called pleurocoels penetrate the lateral sides of the centrum and neural arch. These pleurocoels lead into chambers within the vertebrae that may have contained air. The neural arch of most dinosaur dorsal vertebrae has a tall or thick neural spine where back muscles attached. The ribs connected at lateral processes on the dorsal vertebrae. The ribs of most dinosaurs look alike, so they are difficult or impossible to identify to species. Armored dinosaurs have massive ribs that are often fused to the dorsal vertebrae.

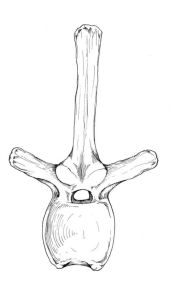

Hadrosaur dorsal vertebra in lateral view

The sacral vertebrae are those upon which the pelvic elements attach. Dinosaurs have two or more sacral vertebrae. In adult dinosaurs, the sacral vertebrae are fused, or locked together, in a tight unit called a sacrum. From each sacral vertebra, a bladelike sacral rib extends outward to connect to the ilium, the main bone

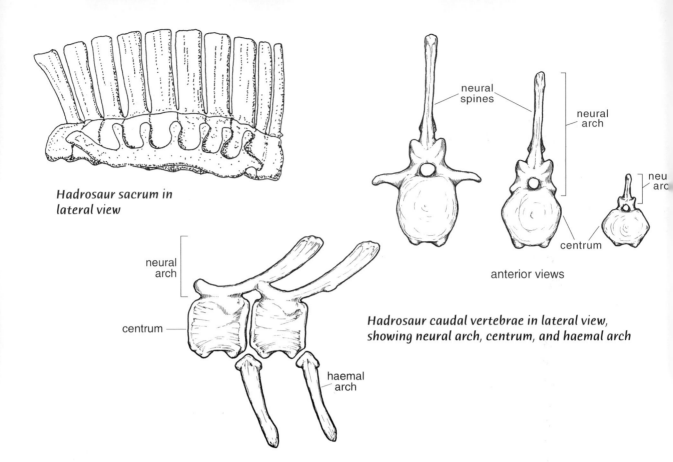

Hadrosaur sacrum in lateral view

neural spines

neural arch

neu arc

centrum

anterior views

centrum

neural arch

haemal arch

Hadrosaur caudal vertebrae in lateral view, showing neural arch, centrum, and haemal arch

of the pelvic girdle. In all dinosaurs, the holes that pass through the neural arches of the sacral vertebrae hold a part of the spinal cord, which was enlarged. Some paleontologists have called this enlarged area of the spinal cord an "extra brain," but it wasn't a brain at all. It was an expanded area of the spinal cord that helped relay information to the brain. Part of that expanded area may have stored other tissues, as it does in animals today.

The bones that make up the tail of all vertebrates are caudal vertebrae. The number of caudal vertebrae in a dinosaur can vary from just a few to more than fifty. Most caudal vertebrae consist of three parts: the centrum, the neural arch, and a lower arch called the haemal arch or better known as the chevron bone. Like the dorsals and sacrals, the caudal vertebrae were amphiplatyan, which means the centrum has flat surfaces, anterior and posterior. The anterior caudal vertebrae of duck-bill dinosaurs, or hadrosaurs, have transverse processes, but the middle and posterior ones do not. Caudal vertebrae can vary considerably in shape within an animal and between different species. However, the anterior caudal

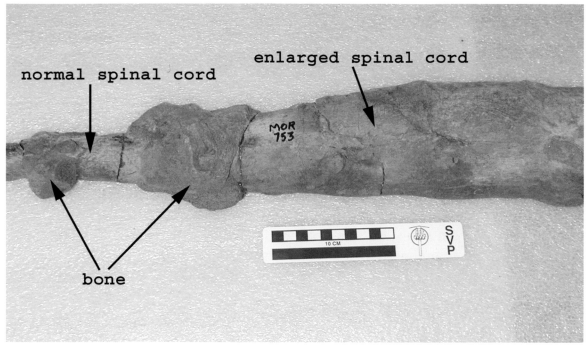

Cast of the neural canal, or spinal cord, from Edmontosaurus annectens, showing the enlarged area in the sacrum

Tail region of Brachylophosaurus showing the ossified tendons that extend down the back and tail

vertebrae of ornithischians are generally round and thin from front to back. Caudal vertebrae from the middle of the tail are generally boxy. The centrum is about as long as it is high and may or may not have transverse processes. The posterior tail vertebrae have long centra. Caudal vertebrae from the ends of the tail of theropod dinosaurs are long and slender, with posterior zygapophyses on the ends of long posterior processes.

On the sides of the neural spines of most ornithischian dinosaur tails are long, thin, rod-shaped bones called ossified tendons. Researchers believe these bones helped hold the tails of these dinosaurs off the ground. Fragments of ossified tendons are common in most sediments containing dinosaur fossils. The fragments are about the diameter of a pencil and very straight.

Front Leg

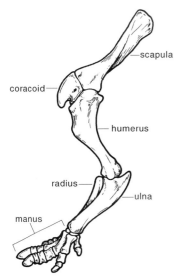

Left front leg of
Leptoceratops

The front legs of any animal are called the pectoral limbs. The bones that connect the pectoral limb to the body are collectively called the pectoral girdle. In a dinosaur, the pectoral girdle includes two bones: the scapula and coracoid. The scapula, commonly known as the shoulder blade, connects with the coracoid to form the shoulder socket in which the humerus, the proximal bone of the arm, sits. The scapula of most dinosaurs is long and flat, and ligaments connect it to the ribs. The coracoid is a circular bone with a hole through it.

Between the pectoral girdles and the lower ends of the ribs are the bones of the sternum and other medial elements. The sternals of most dinosaurs are paired elements. They connect anteriorly to the coracoid bones and posteriorly to either the ribs or the gastralia. Gastralia are riblike bones that form a bony protection for the internal organs on the ventral side of the body. Another bone in some dinosaurs, particularly most of the meat eaters and all the birds, is the furcula, or wishbone. This is one of the features advanced theropod dinosaurs shared with birds. The furcula, as seen in tyrannosaurs and dromaeosaurs, is a V-shaped bone that represents the fusion of the right and left clavicles, or collarbones. As in birds, they connected to the junction of the scapula and coracoid.

The front leg consists of the humerus, ulna, radius, carpals, metacarpals, and manus phalanges. The humerus fits into the

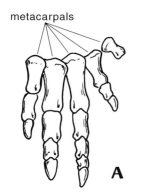

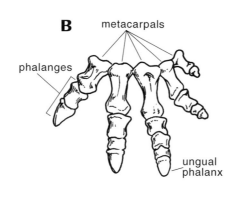

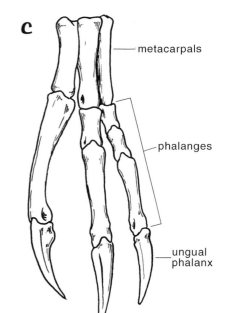

Front feet, or manus, of three dinosaurs: A, Orodromeus; B, Leptoceratops; C, Ornithomimus

shoulder at its proximal end and connects to the ulna and radius at the elbow. Both the proximal and distal ends of the humerus are round in cross section, whereas the proximal and distal ends of the ulna and radius are typically flat. Two rows of small bones called the carpals, or wrist bones, connect to the distal end of the ulna and radius. The carpals connect to the proximal ends of the metacarpals, which make up the bones of the front foot or hand, which we call the manus. Phalanges are the bones that make the fingers or toes of the front foot. The last digit of each finger is called the ungual phalanx. In most dinosaurs, the ungual is a claw or hoof.

Paleontologists number the digits of the manus using Roman numerals from one (I) through five (V) and starting with the thumb. *Orodromeus* and *Leptoceratops* both have five-fingered hands, and *Ornithomimus* has a three-fingered hand—it lost IV and V through evolution, like birds and most other theropods. We number the individual phalanges of each digit starting with 1, the most proximal. The third phalanx on the second digit of a left hand is written as "Lft manus II_3."

The scapula and humerus are often unusual or distinctive enough to distinguish many species of dinosaurs, but most of the other bones of the pectoral girdle and limb are too similar to use for identification. Some metacarpals and phalanges are identifiable to major groups, but not to genus and species.

Hind Leg

The hind leg, or pelvic limb, connects to the pelvic girdle. The pelvic girdle consists of three bones: the ilium (which attaches to the sacrum), the ischium, and the pubis. The pubis is always at the anteroventral, or front, end of the pelvis, and the ischium is at the posteroventral, or back, end, although the pubis bends backward in ornithischians and birds. Together, the three bones of the pelvis fit together to form a large round hole in which the head of the thigh bone, or femur, fits. The hole for the femur is called the acetabulum. The proximal end of the femur has a large, round process that fits into the acetabullum. At the distal end of the femur, two rounded joint surfaces called condyles connect to the proximal end of the tibia and fibula at the knee. The tibia is the largest element. The fibula is usually a thin rod that extends along the lateral surface of the tibia. Dinosaurs didn't have kneecaps. Attached firmly to the lower end of the tibia and fibula are two rows of tarsals that form the ankle. The lower ends of the tarsals meet the metatarsals. These bones connect to the phalanges of the hind foot. As on the front foot, the last phalanges on the hind feet are called unguals and are the bones that had claws or nails. We number the pes as we do the manus. The third phalanx of the first digit of the *Gryposaurus* pes is written as Rt pes I$_3$. Like the front foot (hand) or manus, the hind foot or pes varies considerably between different groups of dinosaurs.

The pelvic bones of dinosaurs are very distinctive and can identify the major group to which the specimens belong. The leg bones are typically less distinctive, and the foot bones are difficult to identify to more than a major group of dinosaurs.

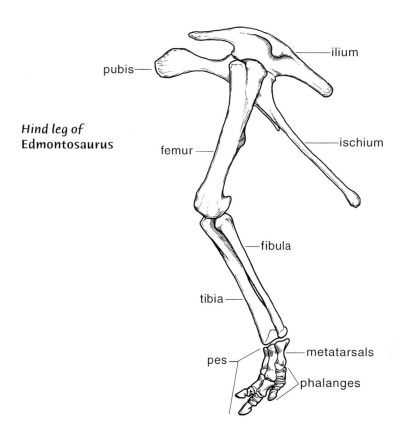

Hind leg of
Edmontosaurus

ilium

pubis

ischium

femur

fibula

tibia

pes

metatarsals

phalanges

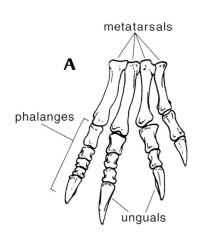

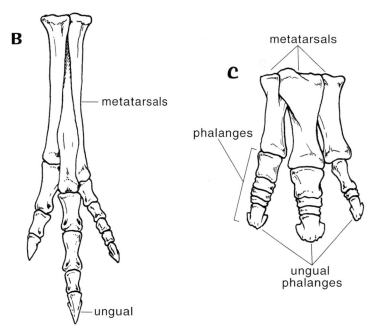

A

metatarsals

phalanges

unguals

B

metatarsals

ungual

C

metatarsals

phalanges

ungual
phalanges

Hind feet, or pes, of three dinosaurs:
 A, right pes of Tenontosaurus;
 B, left pes of Ornithomimus;
 C, right pes of Gryposaurus

Glossary

acetabulum. The hole in the pelvis into which the proximal end of the femur fits.

amphiplatyan. A description of a vertebral centrum that is flat on both the anterior end (toward the head) and posterior end (toward the tail). *Contrast* **opisthocelous** and **procelous.**

anterior. Toward the head end of an animal or bone.

Aptian/Albian Stages. An interval of geological time from 125 million to 100 million years ago during the Cretaceous Period.

articular. Bone in the lower jaw that with the quadrate bone of the skull forms the lower end of the jaw joint. The articular connects to the quadrate bone and prearticular or surangular bones.

articulated. In paleontology, refers to skeletons that remain together with all their bones in place.

atlas. First vertebra in the vertebral column; the vertebra that attaches to the skull and allows it to move up and down.

avian dinosaurs. Birds.

axis. Second vertebra in the vertebral column. The axis attaches to the atlas in front and the third vertebra behind, and allows the skull to move sideways.

basioccipital. Skull bone that connects to the first cervical vertebra of the neck.

bentonite. Volcanic ash that has chemically altered and consists mainly of the clay mineral montmorillonite. Bentonite is also called gumbo.

bonebed. An accumulation of bones that came from more than one individual.

caliche. A carbonate (limestone) mineral that forms in soils and commonly on bones within the soil.

Campanian Stage. An interval of geological time that spanned from 78 million to 70 million years ago during the Cretaceous Period.

carpals. All the bones of the wrist.

cast. Replica of the external details of a bone, shell, or footprint formed when rock or mineral fills a natural mold. *Contrast* **mold.**

caudal vertebrae (caudals). Vertebrae from the tail.

centrosaurine. A group of horned dinosaurs that have short squamosals; for example, *Einiosaurus.*

centrum. The part of the vertebra that is round or spool-shaped. The centrum forms the base of the vertebra and with other centra of the vertebral column forms the central support for the body.

ceratopsians. Horned dinosaurs, a type of ornithischian.

cervical vertebrae (cervicals). Vertebrae from the neck.

chasmosaurine. A group of horned dinosaurs that have long squamosals; for example, *Triceratops.*

chevron. V-shaped bony arch connected to the base of the caudal vertebrae. Also known as the haemal arch. The chevron bones are for connection of tail muscles and protection of nerves and blood vessels.

cladistics. A method of organizing living organisms according to their relation to other organisms. This method assumes that organisms are related by descent.

cladogram. Branching diagram used to display evolutionary relationships of organisms.

coprolite. Fossilized dung or feces.

coracoid. Anterior portion of the shoulder blade that helps to form the socket for the humerus. *See also* **scapula.**

Cretaceous Period. An interval of geological time from 144 million until 64.5 million years ago.

dentary. Bone of the lower jaw that forms the side of the chin.

denticles. The small, raised bumps that make up the serrations of carnivorous dinosaur teeth.

diapsid reptiles (diapsids). Reptiles with two openings (the dorsal and lateral temporal openings) behind the eye orbits. Diapsid reptiles include lizards, snakes, crocodiles, and birds, and all the animals more closely related to them than to turtles.

distal. Farthest from the body's center.

dorsal. Direction toward the back of an animal.

dorsal vertebrae (dorsals). Vertebrae of reptiles that support ribs and lie between the cervical and sacral vertebrae.

evolution. The hypothesis that biological diversity is the result of descent with modification by natural biological processes.

femur. Proximal bone of the hind leg.

fibula. Lesser bone of the lower hind leg. *See also* **tibia.**

foramina (foramen). Small openings in bones through which nerves and vessels pass.

formation. A rock unit consisting of similar sediments and that is traceable over a large enough geographic area that it can be mapped.

furcula. Wishbone, found in many theropod dinosaurs, including birds. The furcula is composed of fused collar bones, or clavicles.

gastralia. Riblike bones that connect to the distal ends of the ribs of theropod dinosaurs to form an articulating bony basket that contains the internal organs.

hadrosaurine. A group of duck-bill dinosaurs that have either no nasal crest or a solid nasal crest; for example, *Edmontosaurus* or *Maiasaura*.

hadrosaurs. Any of the duck-bill dinosaurs.

holotype. The original specimen on which a new species has been based. Also called the type specimen.

humerus. Proximal bone of the arm.

ilium. Primary bone of the pelvis that connects to the sacrum. Together, the ilium, the ischium, and the pubis form the socket for the femur.

ischium. Bone of the pelvis that is posteriorly directed. Together, the ischium, the ilium, and the pubis forms the socket for the femur.

jugal. Skull bone that forms the lower border of the eye orbit and lateral temporal fenestra.

Jurassic Period. An interval of geological time from 208 million to 144 million years ago.

Kimmeridgian Stage. An interval from 150 million to 146 million years ago during the Jurassic Period.

lambeosaurine. A group of duck-bill dinosaurs that have a hollow nasal crest; for example, *Hypacrosaurus.*

lateral. Toward the outside. *Contrast* **medial.**

lateral temporal fenestra. Opening in the side of the skull behind the eye orbit in all diapsid Reptilia, including birds.

Linnaean classification system. A system of organizing living organisms into ordered groups by similarities, predating ideas about their evolutionary relationships.

Maastrichtian Stage. An interval from 70 million to 64.5 million years ago during the Cretaceous Period.

manus. Hand, or front foot.

maxilla. Bone in which the upper cheek teeth reside. In toothed mammals, the maxilla holds the canine teeth, premolars, and molars.

medial. Toward the middle or inside. *Contrast* **lateral.**

Mesozoic Era. An interval of geological time from 230 million until 64.5 million years ago.

metacarpals. Bones of the hand or front foot.

metatarsals. Bones of the hind foot.

microsite. An accumulation of tiny teeth and bones that represent more than one species of animal; typically confined to fragments smaller than the size of an apple.

mold. An impression of the exterior or interior surface of a bone, shell, or footprint. A filled mold would produce a cast. *Contrast* **cast.**

multispecies bonebed. An accumulation of the skeletal remains of more than one species of animal; generally represented by large parts, such as leg bones and vertebrae.

narial opening. Nose hole, or opening in skull where air passes in and out.

nasal. Bone that forms the front end of the narial opening.

neural arch. The portion of the vertebra that covers the spinal chord, supports muscles, and connects to adjacent vertebrae.

nonavian dinosaurs. All dinosaurs except birds.

occipital condyle. The knob on the back of the skull—the posterior end of the basioccipital bone—to which the first vertebra, or atlas, attaches.

opisthocelous. A description of a vertebral centra that is convex toward the head, or anteriorly, and concave toward the tail, or posteriorly. *Contrast* **procelous** and **amphiplatyan.**

orbit. Eye hole, or opening in skull for the eye.

Ornithischia. Group of plant-eating dinosaurs that have a predentary bone.

ornithomimosaurs. The "ostrich dinosaurs."

ornithopods. The ornithischian group of dinosaurs. The premaxillary teeth, or toothless beak, of ornithopods are offset from their maxillary teeth.

ossified tendons. Long, rodlike bones situated against the neural spines of dorsal and caudal vertebrae of most ornithischian dinosaurs. Thought to have helped support the tail and to have been packed in the tail muscles.

pachycephalosaurs. Bone-headed dinosaurs, a type of ornithischian dinosaur, that had a thick bone over the brain.

parietal. The bone or pair of bones that roofs the posterior portion of the brain.

pectoral girdle. Shoulder blade.

pectoral limbs. Front legs.

pelvic girdle. Pelvis.

pelvic limbs. Hind legs.

permineralization. A process of fossilization in which minerals, typically deposited by water, fill the spaces within a bone.

pes. Hind foot.

petrified. The replacement of original bone or wood by another mineral. Also called replacement.

phalanges (*singular:* phalanx). Individual bones of the fingers or toes of either the front or hind foot.

Phanerozoic Eon. An interval of geological time unit from 570 million years ago until the present.

pleurocoels. Openings into the body of the centrum and neural arch of some saurischian dinosaurs.

postcranial. Behind the skull; refers to all the bones of the skeleton except the skull.

posterior. Toward the tail end of an animal or bone.

postorbital. Bone that forms the upper posterior corner of the orbit and the upper anterior border of the lateral temporal fenestra. In ceratopsian dinosaurs, the postorbital forms the orbital horn core.

prearticular. Bone anterior to the articular in the lower jaw.

predentary. Bone only in ornithischian dinosaurs that links the two dentaries together at the front of the mouth.

premaxilla. Bone containing upper incisor teeth or forming the beak.

procelous. A description of a vertebral centra that is concave toward the head, or anteriorly, and convex toward the tail, or posteriorly. *Contrast* **opisthocoelous** and **amphiplatyan.**

process. The end of a bone.

protoceratopsian. Small, primitive member of the ceratopsian family that lacks horns and has a very short neck shield.

proximal. Closest to the body's center.

pubis. Bone of the pelvis that is directed anterior, or toward the head. Together, the pubis, the ilium, and the ischium form the socket for the femur.

quadrate. Bone that unites the cranium with the lower jaw and forms the hinge for the lower jaw. The quadrate forms the posterior side of the lateral temporal fenestra. The upper end of the quadrate fits into a socket in the squamosal, and the lower end of the quadrate connects to the articular bone.

radius. One of two bones that form the lower arm. *See also* **ulna.**

replacement. *See* **petrified.**

Reptilia. A group of vertebrate animals that share a common ancestor and that includes, among many other groups, turtles, lizards, snakes, crocodilians, pterosaurs, dinosaurs, and birds.

rhamphotheca. The horny beak or bill of many reptiles, including birds.

ribs. Bones that connect with the dorsal vertebrae (thoracic vertebrae in mammals) and protect the internal organs.

rostral. The bone that forms the upper beak in ceratopsian dinosaurs.

sacral vertebrae (sacrum). Vertebrae that are fused together and support the pelvic girdle.

sagittal plane. An imaginary vertical plane through the length of an animal from its head to its tail that divides the animal into two similar, mirror-image halves.

Saurischia. Group of dinosaurs with a tri-radiate pelvis, offset thumb, and a neck at least half the length of the trunk. The second finger is always the longest; includes the theropods and sauropods.

sauropods. Group of saurischian dinosaurs with long necks and small heads.

scapula. Posterior, or tailward, portion of the shoulder blade that forms part of the socket for the humerus. *See also* **coracoid.**

sclerotics. Small bones that circle the eyeball of many reptiles, including birds.

Scythian Stage. The earliest stage of the Triassic Period; it lasted from 246 to 241 million years ago.

serrations. Tiny, sharp, bumplike ridges along the anterior and posterior sides of the teeth of carnivores that cut meat like steak knives do. *See also* **denticles.**

squamosal. A bone in the skull of dinosaurs and other animals that unites with the quadrate, supraoccipital, exoccipital, parietal, and postorbital; located on the postero-lateral side of the skull.

stratigraphic level. A point or horizon in a chronological succession of sedimentary rocks that represents a particular sequence of sediment deposition at some point in time.

supratemporal fenestra. Opening in the top of the skull behind the orbits in all diapsid Reptilia, including birds.

surangular. Bone in the lower jaw posterior to the dentary bone.

Synapsida. A group of vertebrates that includes mammals and all animals more closely related to mammals than to reptiles. These animals have a single opening in their cranial elements, behind their orbits.

taphonomy. The study of what happens to organic remains from the time the organism died until its remains are found.

tarsals. Bones of the ankle.

taxon (*plural:* **taxa**). A taxonomic group or entity; a general term used in place of the names of taxonomic groups such as species, genera, and families.

theropods. Meat-eating, bipedal saurischian dinosaurs.

tibia. Primary bone of the lower hind leg. *See also* **fibula.**

transverse plane. An imaginary plane through a skeleton or bone that separates anterior from posterior, or proximal from distal.

transverse process. A lateral bone extension, or rib, on caudal vertebrae centra for the attachment of muscles.

Triassic Period. An interval of geological time from 230 million until 208 million years ago.

thyreophorans. Armored dinosaurs, a type of ornithischian.

type specimen. *See* **holotype.**

ulna. One of two bones of the lower arm segment. The proximal end of the ulna is the elbow. *See also* **radius.**

ungual phalanx. Last bone of each finger or toe; attachment for claws, hoofs, or nails.

ventral. Toward the "belly" of an animal.

vertebra (*plural:* **vertebrae).** Bone of the backbone. Together, the vertebrae form a long group that supports the body and protects the spinal cord.

vertebral column. All the vertebrae that make up the skeleton of a particular animal.

vertebrates. Animals with skeletons of bone or cartilage.

zygapophysis (*plural:* **zygapophyses).** Process on vertebrae that joins vertebrae together. Anterior zygapophyses point upward and inward, and posterior zygapophyses point downward and outward.

Additional Reading

Popular and Scientific* Books that Refer to Montana Dinosaurs

Carpenter, Kenneth. 2000. *Dinosaur Eggs, Embryos and Babies.* Bloomington: Indiana University Press.

*Carpenter, Kenneth, Karl Hirsch, and John R. Horner, eds. 1994. *Dinosaur Eggs and Babies.* Cambridge: Cambridge University Press.

Currie, Phillip J., and Kevin Padian, eds. 1997. *Encyclopedia of Dinosaurs.* San Diego: Academic Press.

Horner, John R., and Don Lessom. 1992. *Digging Up* Tyrannosaurus rex. New York: Crown.

Horner, John R., and Edwin Dobb. 1998. *Dinosaur Lives.* Orlando, Fla.: Harcourt Brace.

Horner, John R., and James Gorman. 1995. *Digging Dinosaurs.* New York: Harper Collins.

———. 1998. *Maia, A Dinosaur Grows Up.* Bozeman: Museum of the Rockies, Montana State University.

Horner, John R., and Don Lessom. 1993. *The Complete* T. rex. New York: Simon and Schuster.

Jaffe, Mark. 2000. *The Gilded Dinosaur.* New York: Crown Publishers.

Johnson, Hope, and John E. Storer. 1974. *A Guide to Vertebrate Fossils from the Age of Dinosaurs.* Drumheller: Provincial Museum of Alberta, Publication 4.

Norell, Mark A., Eugene S. Gaffney, and Lowell Dingus. 1995. *Discovering Dinosaurs.* New York: Alfred A. Knopf.

Paul, Gregory S. 1988. *Predatory Dinosaurs of the World, A Complete Illustrated Guide.* New York: Simon and Schuster.

Wallace, David Rains. 1999. *The Bone Hunters' Revenge.* Boston: Houghton Mifflin.

Warren, Leonard. 1998. *Joseph Leidy, The Last Man Who Knew Everything.* New Haven, Conn: Yale University Press.

*Weishampel, David B., Peter Dodson, and Halska Osmolska, eds. 1990. *The Dinosauria.* Berkeley: University of California Press.

Books about Dinosaurs from Other States

DeCourten, Frank. 1998. *Dinosaurs of Utah.* Salt Lake City: University of Utah Press.

Gallagher, William. 1997. *When Dinosaurs Roamed New Jersey.* New Brunswick, N.J.: Rutgers University Press.

Jacobs, Louis. 1995. *Lone Star Dinosaurs.* Louise Lindsey Merrick Natural Environment Series, 22. Dallas: Texas A&M University Press.

Weishampel, D., and Luther Young. 1996. *Dinosaurs of the East Coast.* Baltimore: Johns Hopkins University Press.

Scientific Publications about Montana Dinosaurs

Archibald, J. D., and L. J. Bryant. 1990. *Differential Cretaceous/Tertiary extinctions of nonmarine vertebrates; evidence from northeastern Montana.* Geological Society of America, Special Paper 247:549–62.

Archibald, J. D., and W. A. Clemens. 1982. Late Cretaceous Extinctions. *American Scientist* 70:377–85.

Baird, D. 1979. The dome-headed dinosaur *Tylosteus ornatus* Leidy 1872 (Reptilia, Ornithischia, Pachycephalosauridae). *Notulae Naturae.* Philadelphia: The Academy of Natural Sciences. 456:1-11.

Bakker, R. T., M. Williams, and P. J. Currie. 1988. *Nanotyrannus*, a new genus of pygmy tyrannosaur from the latest Cretaceous of Montana. *Huntaria* 1(5):1-30.

Barreto, C., R. M. Albrecht, D. E. Bjorling, J. R. Horner, and N. J. Wilsman. 1993. Evidence of the growth plate and the growth of long bones in juvenile dinosaurs. *Science* 262:2020-23.

Barrick, R. E., W. J. Showers, and A. G. Fischer. 1996. Comparison of thermoregulation of four ornithischian dinosaurs and a varanid lizard from the Cretaceous Two Medicine Formation: Evidence from oxygen isotopes. *Palaios* 11:295-305.

Brett-Surman, M. K., and G. S. Paul. 1985. A new family of bird-like dinosaurs linking Laurasia and Gondwanaland. *Journal Vertebrate Paleontology* 5(2):133–38.

Brown, B. 1907. The Hell Creek beds of the Upper Cretaceous of Montana: Their relation to contiguous deposits, with faunal and floral lists and discussion of their correlation. *Bulletin of the American Museum of Natural History* 23:823-845.

——. 1908. The Ankylosauridae, a new family of armored dinosaurs from the Upper Cretaceous. *American Museum Natural History Bulletin* 24:187-201.

——. 1933. A gigantic ceratopsian dinosaur, *Triceratops maximus*, new species. *American Museum Novitates* 649:1-9.

Brown, B., and E. M. Schlaikjer. 1942. The skeleton of *Leptoceratops* with a description of a new species. *American Museum Novitates* 1169:1-15.

Burnham, D. A., K. L. Drestler, and C. J. Linster. 1997. A new specimen of *Velociraptor* (Dinosauria: Theropoda) from the Two Medicine Formation of Montana. In *Dinofest International Philadelphia,* edited by D. L. Wolberg, E. Stump, and G. D. Rosenberg, 73-75. Philadelphia: The Academy of Natural Sciences.

Burnham, D. A., K. L. Drestler, P. J. Currie, R. T. Bakker, Z. Zhou, and J. H. Ostrom. 2000. Remarkable new birdlike dinosaur (Theropoda: *Maniraptora*) from the Upper Cretaceous of Montana. *The University of Kansas Paleontological Contributions* 13:1-14.

Carpenter, K. 1982. Baby dinosaurs from the Late Cretaceous Lance and Hell Creek Formations and a description of a new species of theropod. *Contributions to Geology, University of Wyoming* 20(2):123–34.

——. 1984. Skeletal reconstruction and life restoration of *Sauropelta* (Ankylosauria: Nodosauria) from the Cretaceous of North America. *Canadian Journal of Earth Sciences* 21:1491–98.

——. 1990. Ankylosaur systematics: example using *Panoplosaurus* and *Edmontonia* (Ankylosauria: Nodosauridae). Pp. 281-298 In *Dinosaur Systematics, Approaches and Perspectives,* edited by K. Carpenter and P. J. Currie. Cambridge: Cambridge University Press.

Carpenter, K., and K. Alf. 1994. Global distribution of dinosaur eggs, nests and babies. Pp. 15-30 In *Dinosaur Eggs and Babies,* edited by K. Carpenter, K. Hirsch, and J. R. Horner. Cambridge: Cambridge University Press.

Carpenter, K., and B. Breithaupt. 1986. Latest Cretaceous occurrence of nodosaurid ankylosaurs (Dinosauria, Ornithischia) in western North America and the gradual extinction of the dinosaurs. *Journal of Vertebrate Paleontology* 6(3):251–57.

Carr, T. D. 1999. Craniofacial ontogeny in Tyrannosauridae (Dinosauria, Coelurosauria). *Journal of Vertebrate Paleontology* 19(3):497–520.

Chinnery, B. J., and D. B. Weishampel. 1998. *Montanaceratops cerorhynchus* (Dinosauria: Ceratopsia) and relationships among basal neoceratopsians. *Journal of Vertebrate Paleontology* 18(3):569–85.

Cobabe, E. A., and D. E. Fastovski. 1987. *Ugorosaurus olsoni*, a new ceratopsian (Reptilia: Ornithischia) from the Hell Creek Formation of Montana. *Journal of Paleontology* 61(1):148–54.

Coombs, W. P. Jr. 1995. Ankylosaurian tail clubs of middle Campanian to early Maastrichtian age from western North America, with description of a tiny tail club from Alberta and discussion of tail orientation and tail club function. *Canadian Journal of Earth Sciences* 32:902–12.

Cope, E. D. 1876. Description of some vertebrate remains from the Fort Union beds of Montana. *Proceedings of the Academy of Natural Sciences, Philadelphia 1876.* 248–61.

——. 1877a. On some extinct reptiles and batrachia from the Judith River and Fox Hills beds of Montana. *Proceedings of the Academy of Natural Sciences, Philadelphia 1876.* 340–59.

——. 1877b. The discovery of *Laelaps* in Montana. *American Naturalist* 12:311.

——. 1888. A horned dinosaurian reptile. *American Naturalist* 22:1108–9.

Creisler, B. S. 1992. Why *Monoclonius* Cope was not named for its horn: the etymologies of Cope's dinosaurs. *Journal of Vertebrate Paleontology* 12(3):313–17.

Currie, P. J., J. K. Rigby Jr., and R. E. Sloan. 1990. Theropod teeth from the Judith River Formation of southern Alberta, Canada. In *Dinosaur Systematics, Approaches and Perpectives,* edited by K. Carpenter and P. J. Currie, 107–25. Cambridge: Cambridge University Press.

Dodson, P. 1984. Small Judithian ceratopsids, Montana and Alberta. In *3rd Symposium of Mesozoic Terrestrial Ecosystems Short Papers,* edited by W-E Reif and F. Westphal, 73–78, Tübingen, Germany: Attempto Verlag.

———. 1986. *Avaceratops lammersi*: a new ceratopsid from the Judith River Formation of Montana. *Proceedings of the Academy of Natural Sciences, Philadelphia* 138(2):305–17.

———. 1996. *The Horned Dinosaurs, A Natural History.* Princeton: Princeton University Press.

Douglas, E. 1902. Dinosaurs in the Ft. Pierre Shale and underlying beds in Montana. *Science* 15:31–32.

Erickson, B. R. 1960. Prospecting for dinosaurs in Montana. *Museum Observer* 1:2–24.

———. 1966. Mounted skeleton of *Triceratops prorsus* in the Science Museum. *Science Museum Publications* 1:1–16.

Erickson, G. M. 1995. Split carinae on tyrannosaurid teeth and implications of their development. *Journal of Vertebrate Paleontology* 15(2):268–74.

———. 1999. Breathing life into *Tyrannosaurus rex. Scientific American* 281(3):44–49.

Erickson, G. M., and K. L. Olson. 1996. Bite marks attributable to *Tyrannosaurus rex*: preliminary description and implications. *Journal of Vertebrate Paleontology* 16(1):175–78.

Erickson, G. M., S. D. van Kirk, J. Su, M. E. Levenston, W. E. Caler, and D. R. Carter. 1996. Bite-force estimation for *Tyrannosaurus rex* from tooth-marked bones. *Nature* 382:706–8.

Farlow, J. O., M. B. Smith, and J. M. Robinson. 1995. Body mass, bone "strength indicator," and cursorial potential of *Tyrannosaurus rex. Journal of Vertebrate Paleontology* 15(4):713–25.

Fiorillo, A. R. 1987. Significance of juvenile dinosaurs from Careless Creek Quarry (Judith River Formation), Wheatland County, Montana. In *Fourth Symposium Mesozoic Terrestrial Ecosystems,* edited by P. J. Currie and E. H. Koster, 88–95. Drumheller, Alberta: Tyrrell Museum of Palaeontology.

———. 1989. The vertebrate fauna of the Judith River Formation (Late Cretaceous) of Wheatland and Golden Counties, Montana. *The Mosasaur* 4:127–42.

———. 1990. The first occurrence of hadrosaur (Dinosauria) remains from the marine Claggett Formation, Late Cretaceous of south-central Montana. *Journal of Vertebrate Paleontology* 10(4):515–17.

———. 1991. Taphonomy and depositional setting of Careless Creek Quarry (Judith River Formation, Wheatland County, Montana, U.S.A. *Palaeogeography, Palaeoclimatology, Palaeoecology* 81:281–311.

Fiorillo, A. R., and P. J. Currie. 1994. Theropod teeth from the Judith River Formation (Upper Cretaceous) of south-central Montana. *Journal of Vertebrate Paleontology* 14:74–80.

Forester, C. A. 1990. Evidence for juvenile groups in the ornithopod dinosaur *Tenontosaurus tilletti* Ostrom. *Journal of Paleontology* 64(1):164–65.

———. 1996. New information on the skull of *Triceratops*. *Journal of Vertebrate Paleontology* 16(2):246–58.

———. 1996. Species resolution in *Triceratops*: cladistic and morphometric approaches. *Journal of Vertebrate Paleontology* 16(2):259–70.

Foster, Mike. 1994. *Strange Genius: The Life of Ferdinand Vandeveer Hayden*. Niwot, Colo.: Roberts Rinehart Publishers.

Galton, P. M., and H.-D. Sues. 1983. New data on pachycephalosaurid dinosaurs (Reptilia: Ornithischia) from North America. *Canadian Journal of Earth Sciences* 20(3):462–72.

Giffin, E. B., D. L. Gabriel, and R. E. Johnson. 1987. A new pachycephalosaurid skull (Ornithischia) from the Cretaceous Hell Creek Formation of Montana. *Journal of Vertebrate Paleontology* 7(4):398–407.

Gilmore, C. W. 1917. Brachyceratops, *a ceratopsian from the Two Medicine Formation of Montana, with notes on associated fossil reptiles*. U.S. Geological Survey Professional Paper 103:1-45.

———. 1922. The smallest known horned dinosaur, *Brachyceratops*. *Proceedings of the U.S. National Museum* 61(3):1–4.

———. 1929. Hunting dinosaurs in Montana. *Exploration and Field-Work of the Smithsonian Institution in 1928*, 7–12.

———. 1930. On dinosaurian reptiles from the Two Medicine Formation of Montana. *Proceedings of the U.S. National Museum* 77(16):1–39.

———. 1933. On dinosaurian reptiles from the Two Medicine Formation of Montana. *Proceedings of the U.S. National Museum* 77:1–38.

———. 1935. Fossil hunting in Montana and Wyoming. *Smithsonian Institution, Smithsonian Explorations* 1935:1–4.

———. 1937. On the detailed skull structure of a crested hadrosaurian dinosaur. *Proceedings of the U.S. National Museum* 84:481–91.

———. 1939. Ceratopsian dinosaurs from the Two Medicine Formation, Upper Cretaceous of Montana. *Proceedings of the U.S. National Museum* 87:1–18.

———. 1946. A new carnivorous dinosaur from the Lance Formation of Montana. *Smithsonian Institution Miscellaneous Collections* 106(13):1–19.

Goodwin, M. B. 1990. Morphometric landmarks of pachycephalosaurid cranial material from the Judith River Formation of northcentral Montana. In *Dinosaur Systematics, Approaches and Perspectives*, edited by K. Carpenter and P. J. Currie, 189–201. Cambridge: Cambridge University Press.

Goodwin, M. B., E. A. Buchholtz, and R. E. Johnson. 1998. Cranial anatomy and diagnosis of *Stygimoloch spinifer* (Ornithischia: Pachycephalosauria) with comments on cranial display structures in agnostic behavior. *Journal of Vertebrate Paleontology* 18(2):363–75.

Hatcher, J. B. 1907. The Ceratopsia. *Monographs of the United States Geological Survey* XLIX:300 pp.

Hirsch, K. F., and B. Quinn. 1990. Eggs and eggshell fragments from the Upper Cretaceous Two Medicine Formation of Montana. *Journal of Vertebrate Paleontology* 10:491–511.

Horner, J. R. 1979. Upper Cretaceous dinosaurs from the Bearpaw Shale (marine) of south-central Montana with a checklist of Upper Cretaceous dinosaur remains from marine sediments in North America. *Journal of Paleontology* 53(3):566–77.

——. 1982. Evidence of colonial nesting and site fidelity among ornithischian dinosaurs. *Nature* 297:675–76.

——. 1983. Cranial osteology and morphology of the type specimen of *Maiasaura peeblesorum* (Ornithischia; Hadrosauridae), with discussion of its phylogenetic position. *Journal of Vertebrate Paleontology* 3(1):29–38.

——. 1984. Three ecologically distinct vertebrate faunal communities from the Late Cretaceous Two Medicine Formation of Montana, with discussion of evolutionary pressures induced by interior seaway fluctuations. *Montana Geological Society, 1984 Field Conference and Symposium Guidebook,* 299–303.

——. 1984. A segmented epidermal tail frill in a species of hadrosaurian dinosaur. *Journal of Paleontology* 58:270–71.

——. 1984. The nesting behavior of dinosaurs. *Scientific American* 250(4):130–37.

——. 1987. Ecological and behavioral implications derived from a dinosaur nesting site. In *Dinosaurs Past and Present, Vol II.,* edited by S. Czerkas and E. C. Olson, 50–63. Seattle: University of Washington Press.

——. 1988. A new hadrosaur (Reptilia, Ornithischia) from the Upper Cretaceous Judith River Formation of Montana. *Journal of Vertebrate Paleontology* 8(3):314–21.

——. 1989. The Mesozoic terrestrial ecosystems of Montana. *Montana Geological Society, 1989 Field Conference and Symposium Guidebook,* 153–62.

——. 1992. *Cranial morphology of* Prosaurolophus *(Ornithischia: Hadrosauridae) with descriptions of two new hadrosaurid species and an evaluation of hadrosaurid phylogenetic relationships.* Museum of the Rockies Occasional Paper 2:1–119.

——. 1994. Comparative taphonomy of some dinosaur and extant bird colonial nesting grounds. In *Dinosaur Eggs and Babies,* edited by K. Carpenter, K. Hirsch, and J. R. Horner, 116–23. Cambridge: Cambridge University Press.

——. 1997. Rare preservation of an incompletely ossified fossil embryo. *Journal of Vertebrate Paleontology* 17(2):431–34.

——. 1999. Egg Clutches and embryos of two hadrosaurian dinosaurs. *Journal of Vertebrate Paleontology* 19(4):607–11.

———. 2000. Dinosaur reproduction and parenting. *Annual Reviews of Earth and Planetary Sciences* 28:19–45.

Horner, J. R., and V. Clouse. 1998. An undisturbed clutch of hadrosaur eggs from the Judith River Formation of Montana. *1st International Meeting on Dinosaur Paleobiology, Museu Nacional de Historia Natural, Universidade de Lisboa, 1998, Abstracts and Program.* 22–25.

Horner, J. R., and P. J. Currie. 1994. Embryonic and neonatal morphology of a new species of *Hypacrosaurus* (Ornithischia, Lambeosauridae) from Montana and Alberta. In *Dinosaur Eggs and Babies,* edited by K. Carpenter, K. Hirsch, and J. R. Horner, 312–36. Cambridge: Cambridge University Press.

Horner, J. R., and R. Makela. 1979. Nest of juveniles provides evidence of family structure among dinosaurs. *Nature* 282:296–98.

Horner, J. R., A. de Ricqlès, and K. Padian. 1999. Variation in dinosaur skeletochronology indicators: implications for age assessment and physiology. *Paleobiology* 25(3):295–304.

———. 2000. Long bone histology of the hadrosaurid dinosaur *Maiasaura peeblesorum:* growth dynamics and physiology based on an ontogenetic series of skeletal elements. *Journal of Vertebrate Paleontology* 20(1):109–23.

Horner, J. R., D. J. Varricchio, and M. Goodwin. 1992. Marine transgressions and the evolution of Cretaceous dinosaurs. *Nature* 358:59–61.

Horner, J. R., and D. B. Weishampel. 1988. A comparative embryological study of two ornithischian dinosaurs. *Nature* 332:256–57.

———. 1996. A comparative embryological study of two ornithischian dinosaurs, a correction. *Nature* 383:103.

Jepson, G. L. 1931. Dinosaur egg shell fragments from Montana. *Science* 73:12–13.

Leidy, J. 1856. Notices of remains of extinct reptiles and fishes, discovered by Dr. F. V. Hayden in the badlands of the Judith River, Nebraska Territory. *Proceedings of the Academy Natural Sciences, Philadelphia 1856* 8:72–73.

———. 1859. Extinct vertebrata from the Judith River and great lignite formations of Nebraska. *American Philosophical Transactions* 11:139–54.

———. 1873. *Contributions to the extinct vertebrate fauna of the Western Territories.* Report of the United Geological Survey of the Territories, Government Printing Office, Washington, D.C.

Lofgren, D. L., C. L. Hotton, and A. D. Runkel. 1990. Reworking of Cretaceous dinosaurs into Paleocene channel deposits, Upper Hell Creek Formation, Montana. *Geology* 18:874–77.

Lull, R. S. 1903. Skull of *Triceratops serratus. Bulletin of the American Museum of Natural History* 19(30):685–95.

Lull, R. S., and N. E. Wright. 1942. *Hadrosaurian dinosaurs of North America.* Geological Society of America, Special Paper 40:1–242.

MacDonald, J. R. 1967. The *Tyrannosaurus* research goes on. *Los Angeles County Museum Natural History Quarterly* 5:12–14.

Makovicky, P. J., and H.-D. Sues. 1998. Anatomy and phylogenetic relationships of the theropod dinosaur *Microvenator celer* from the Lower Cretaceous of Montana. *American Museum Novitates* 3240:1–28.

Marsh, O. C. 1888. A new family of horned dinosaurs from the Cretaceous. *American Journal of Science* 36:447–78.

——. 1889. Notice of new American dinosaurs. *American Journal of Science* 37:331–36.

——. 1890. Description of new dinosaurian reptiles. *American Journal of Science* 39:81–86.

Maxwell, D. 1995. Taphonomy and paleobiological implications of *Tenontosaurus-Deinonychus* associations. *Journal of Vertebrate Paleontology* 15(4):707–12.

Maxwell, D., and J. R. Horner. 1994. Neonate dinosaurian remains and dinosaurian eggshell from the Cloverly Formation, Montana. *Journal of Vertebrate Paleontology* 14(1):143–46.

McMannis, W. J. 1957. The Livingston Formation. *Billings Geological Society, 8th Annual Field Conference Guidebook,* 80–84.

Molnar, R. 1978. A new theropod dinosaur from the Upper Cretaceous of central Montana. *Journal of Paleontology* 52:73–82.

——. 1979. Correction: location of dromaeosaurid. *Journal of Paleontology* 53:1256.

——. 1980. An albertosaur from the Hell Creek Formation of Montana. *Journal of Paleontology* 54:102–8.

——. 1991. The cranial morphology of *Tyrannosaurus rex*. *Palaeontographica* 217:137–76.

Morris, W. J. 1976. Hypsilophodont dinosaurs: a new species and comments on their systematics. In *Athlon, Royal Ontario Museum of Life Sciences, Miscellaneous Publications,* edited by C. S. Churcher, 93–113.

——. 1978. *Hypacrosaurus altispinus?* Brown from the Two Medicine Formation, Montana, a taxonomically indeterminate specimen. *Journal of Paleontology* 52:200–205.

Osborn, H. F. 1905. *Tyrannosaurus* and other Cretaceous carnivorous dinosaurs. *Bulletin American Museum Natural History* 21(14):259–65.

——. 1906. *Tyrannosaurus*, Upper Cretaceous carnivorous dinosaur (second communication). *Bulletin American Museum Natural History* 22(16):281–96.

——. 1912. Crania of *Tyrannosaurus* and *Allosaurus*. *American Museum of Natural History Memoirs* (new series) 1 (Part I):30 pp.

——. 1933. Mounted skeleton of *Triceratops elatus*. *American Museum Novitates* 564:1–14.

Ostrom, J. H. 1964. The systematic position of *Hadrosaurus* (*Ceratops*) *paucidens* (Marsh). *Journal of Paleontology* 38:130–34.

——. 1969. A new theropod dinosaur from the Lower Cretaceous of Montana. *Postilla* 128:1–17.

———. 1969. Osteology of *Deinonychus antirrhopus*, an unusual theropod from the Lower Cretaceous of Montana. *Peabody Museum Natural History, Yale University Bulletin* 30:1–165.

———. 1970. Stratigraphy and paleontology of the Cloverly Formation (Lower Cretaceous) of the Big Horn Basin area of Wyoming and Montana. *Peabody Museum Natural History, Yale University Bulletin* 35:1–234.

———. 1974. The pectoral girdle and forelimb function of *Deinonychus* (Reptilia, Saurischia): A correction. *Postilla* 165:1–11.

Ostrom, J. H., and P. Wellnhofer. The Munich specimen of *Triceratops* with a revision of the genus. *Zittliana* 14:111–58.

Penkalski, P., and P. Dodson. 1999. The morphology and systematics of *Avaceratops*, a primitive horned dinosaur from the Judith River Formation (Late Campanian) of Montana with the description of a second skull. *Journal of Vertebrate Paleontology* 19(4):692–711.

Perry, E. S. 1962. Montana in the Geologic Past. *Montana Bureau of Mines and Geology Bulletin* 26:1–78.

Rigby, J. K. Jr., K. R. Newman, J. Smit, S. Van der Kars, R. E. Sloan, and J. K. Rigby. 1987. Dinosaurs from the Paleocene part of the Hell Creek Formation, McCone County, Montana. *Palaios* 2:296–302.

Rogers, R. R. 1990. Taphonomy of three dinosaur bone beds in the Upper Cretaceous Two Medicine Formation of northwestern Montana: Evidence for drought-related mortality. *Palaios* 5:394–413.

———. 1992. Non-marine borings in dinosaur bones from the Upper Cretaceous Two Medicine Formation, northwestern Montana. *Journal of Vertebrate Paleontology* 12(4):528–31.

Rohrer, W. L., and Konizeski, R., 1960. On the occurrence of *Edmontosaurus* in the Hell Creek Formation of Montana. *Journal of Paleontology* 34(3):464–66.

Russell, D. A. 1982. *A paleontological consensus of the extinction of the dinosaurs?* Geological Society of America Special Paper 190:401–5.

———. 1982. The mass extinctions of the Late Mesozoic. *Scientific American* 246(1):58–65.

Russell, L. S. 1964. Cretaceous non-marine faunas of northwestern North America. *Royal Ontario Museum Life Sciences Contribution* 61:1–24.

———. 1968. A dinosaur bone from the Willow Creek beds in Montana. *Canadian Journal of Earth Sciences* 5:327–29.

Sahni, A. 1972. The vertebrate fauna of the Judith River Formation, Montana. *Bulletin of the American Museum of Natural History* 147:321–412.

Sampson, S. D. 1995. Two new horned dinosaurs from the Upper Cretaceous Two Medicine Formation of Montana; with a phylogenetic analysis of the Centrosaurinae (Ornithischia: Ceratopsidae). *Journal of Vertebrate Paleontology* 15(4):743–60.

——. 1997. Craniofacial ontogeny in centrosaurine dinosaurs (Ornithischia: Ceratopsidae): taxonomic and behavioral implications. *Zoological Journal of the Linnean Society* 121:293–337.

Schwietzer, M. H., C. Johnson, T. G. Zocco, J. R. Horner, and J. R. Starkey. 1997. Preservation of biomolecules in cancellous bone of *Tyrannosaurus rex*. *Journal of Vertebrate Paleontology* 17(2):349–59.

Schwietzer, M. H., and J. R. Horner. 1999. Intervascular microstructures in trabecular bone tissues of *Tyrannnosaurus rex*. *Annals Palèontologie* 85(3):179–92.

Schwietzer, M. H., M. Marshal, K. Caron, D. S. Bohle, S. C. Busse, E. V. Arnold, D. Barnard, J. R. Horner, and J. R. Starkey. 1997. Heme compounds in dinosaur trabecular bone. *Proceedings of the National Academy of Sciences* 94:6291–96.

Sheehan, P. M., D. E. Fastovsky, R. G. Hoffmann, C. B. Berghaus, and D. L. Gabriel. 1991. Sudden extinction of the dinosaurs: Latest Cretaceous, Upper Great Plains, USA. *Science* 254:835–39.

Sloan, R. E. 1976. The ecology of dinosaur extinction. *Athlon, Royal Ontario Museum of Life Sciences, Miscellaneous Publications*. Royal Ontario Museum, Misc. Publications 1976:13454.

Sloan, R. E., J. K. Rigby, L. M. Van Valen, and D. Gabriel. 1986. Gradual dinosaur extinction and simutaneous ungulate radiation in the Hell Creek Formation. *Science* 232:629–33.

Stanton, T. W., and J. B. Hatcher. 1905. Geology and paleontology of the Judith River beds. *Bulletin of the U.S. Geological Survey* 257:1–128.

Sues, H.-D. 1980. Anatomy and relationships of a new hypsilophodontid dinosaur from the Lower Cretaceous of North America. *Palaeontographica* 169:51–72.

Sues, H.-D., and P. M. Galton. 1987. Anatomy and classification of the North American Pachycephalosauria (Dinosauria: Ornithischia). *Palaeontographica* 198:1–40.

Truman, C. N. 1999. Rare earth element geochemistry and taphonomy of terrestrial vertebrate assemblages. *Palaios* 14(6):555–68.

Turner, C. E., and F. Peterson. 1999. Biostratigraphy of dinosaurs in the Upper Jurassic Morrison Formation of the Western Interior, U.S.A. In *Vertebrate Paleontology in Utah*, edited by D. D. Gillette. Utah Geological Survey Miscellaneous Publications 99-1:77–114.

Van Valen, L., and R. E. Sloan. 1977. Ecology and the extinction of the dinosaurs. *Evolutionary Theory* 2:37–64.

Varricchio, D. J. 1993. Bone microstructure of the Upper Cretaceous dinosaur *Troodon formosus*. *Journal of Vertebrate Paleontology* 13:99–104.

——. 1995. Taphonomy of Jack's Birthday Site, a diverse dinosaur bonebed from the Upper Cretaceous Two Medicine Formation of Montana. *Palaeogeography, Palaeoclimatology, Palaeoecology* 114:297–323.

Varricchio, D. J., and J. R. Horner. 1993. Hadrosaurid and lambeosaurid bone beds from the Upper Cretaceous Two Medicine Formation of Montana: taphonomic and biologic implications. *Canadian Journal of Earth Sciences* 30:997–1006.

Varricchio, D. J., F. Jackson, J. J. Borkowski, and J. R. Horner. 1997. Nest and egg clutches of the dinosaur *Troodon formosus* and the evolution of avian reproductive traits. *Nature* 385:247–50.

Varricchio, D. J., F. Jackson, and C. N. Truman. 1999. A nesting trace with eggs for the Cretaceous theropod dinosaur *Troodon formosus*. *Journal of Vertebrate Paleontology* 19(1):91–100.

Wall, W. P., and P. M. Galton. 1979. Notes on pachycephalosaurid dinosaurs (Reptilia: Ornithischia) from North America, with comments on their status as ornithopods. *Canadian Journal of Earth Sciences* 16(6):1176–86.

Weishampel, D. B. 1990. Dinosaurian distribution. In *The Dinosauria*, D. B. Weishampel, P. Dodson, and H. Osmólska, 63–139. Berkeley: University of California Press.

Weishampel, D. B., and J. R. Horner. 1987. *Dinosaurs, habitat bottlenecks, and the St. Mary River Formation.* Occasional Paper of the Tyrrell Museum of Palaeontology (Drumheller, Alberta) 3:224–29.

White, P. D., D. E. Fastovsky, and P. M. Sheehan. 1998. Taphonomy and suggested structure of the dinosaurian assemblage of the Hell Creek Formation (Maastrichtian), eastern Montana and western North Dakota. *Palaios* 13(1):41–50.

Williams, M. E. 1994. Catastrophic versus noncatastrophic extinction of the dinosaurs: Testing, falsifiability, and the burden of proof. *Journal of Paleontology* 68(2):183–90.

Index

abbreviations for museums and land management agencies, 83

Achelousaurus horneri, 56-7, 71, 109

Albertosaurus, 71, 113; *libratus,* 122-23

allosaurids: in Dinosauria cladogram, 15

Allosaurus fragilis, 63, 91-92

amateur paleontology, 3-4

American Museum of Natural History, 48

anatomy, dinosaur skeletal, 155-69; anatomical directions, 155-56; front leg, 166-67; hind leg, 168-69; manus, 167; pes 169; skull 157-59; tail, 164-66; teeth, 160-62; vertebrae 162-4

Anatotitan, 52; *copei,* 134

angiosperms (flowering plants), 63

Ankylosauridae, 110-11, 134-35

Ankylosaurus, 2, 15; *magniventris,* 134-35

Apatosaurus, 15, 63, 89-90

Aptian/Albian Stage (Cretaceous), 63-67; Cloverly Formation, 63, 66, 93-100; geological time scale, 17; importance, 18; Kootenai Formation, 63, 66; Montana depositional features map, 63; ornithischian dinosaurs from, 93-96; saurischian dinosaurs from, 96-100

argon 40—argon 39 radiometric dating, 19

armored dinosaurs. *See* Ankylosauridae

Avaceratops lammersi, 2, 71, 121-22; first specimen discovered, 55

baenids, 117

Baird, Don, 82

Bakker, Bob, 53

Bambiraptor, 83; *feinbergi,* 115

Barosaurus, 63

Bearpaw Formation (marine), 24, 76; in stratigraphic section, 23, 24; on map of Montana Maastrichtian depositional features, 76

Big Sandy, Montana, 69

Billings, Montana, 69

Billman Creek Formation, 77, 79

birds: in Dinosauria cladogram, 15; relationship to dinosaurs, 13-14

Black Coulee National Wildlife Refuge, 69

Blackfeet Nation, 56, 57, 71, 79

bonebeds, 32; multispecies, 33-34

Boychuk, Larry, 59

Brachyceratops montanensis, 52, 110

Brachylophosaurus, 31, 71; *canadensis,* 120-21; *goodwini,* 120-21

Brandvold, Marion, 56

Bridger, Montana, 61

Brockton, Montana, 25

Brown, Barnum, 48-51

Browning, Montana: geological surroundings of, 24; in stratigraphic section, 24

Buganosaura infernalis, 128-29

Bureau of Land Management, 66

Cambrian Period: geological time scale, 17

Campanian Stage: dinosaurs from, 100-124; geological time scale, 17; importance, 18; Judith River Formation, 70, 74-75, 117-24; on map of Montana depositional features, 69

"Camposaur" Quarry, map of, 33

Camptosaurus, 63, 87

Carter County Museum, 52

cataloging fossils, 148-50

Cenozoic Era: geological time scale, 17

cephalopods, 69

Ceratodus (fish), 66; *frazieri* (lungfish), 99

Ceratops montanus, 46

Ceratopsidae, 108-10, 130-32

Chester, Montana, 25

chimaera, 11

Choteau, Montana, 19, 71

Chugwater Formation: in stratigraphic section, 23; on map of Montana's Triassic depostional features, 61

cladistics, 13, 14

Claggett Formation (marine): exposure, 24, 25; in stratigraphic section, 23; marine fauna from, 69

classification of dinosaurs, 13

Clouse, Vicki, 58

Cloverly Formation (Early Cretaceous): description of, 66; dinosaurs from, 93–100; exploration of, 53, 55; fossils of, 66; gizzard stones in, 43; importance of, 22; in stratigraphic section, 23; on map of Montana's Mesozoic geology, 22; ornithischian dinosaurs of, 93–96; saurischian dinosaurs of, 97–100

Colorado Group: exposure of, 24; in stratigraphic section, 23, 24

commercial fossil collecting, 10, 83

Compsemys (turtle), 117

concretions, 141–43

conifers, 63, 66

Cope, Edward Drinker, 45, 46

coprolite, 35, 36, 42, 43

Cretaceous Period: 16, 63–80; Cloverly Formation (Early Cretaceous) dinosaurs, 93–99; Cloverly Formation nondinosaurian fossils, 99–100; geological time scale, 17; Hell Creek Formation aves and vertebrates, 138–40; Hell Creek Formation (Late Cretaceous) dinosaurs, 127–38; Judith River Formation (Late Cretaceous) dinosaurs, 118–23; Judith River Formation non-vertebrates, 124; Livingston Group (Late Cretaceous) dinosaurs, 125–26; maps of Montana's depositional features during, 63, 69, 70, 76, 77; Saint Mary River Formation (Late Cretaceous) 126–27; Two Medicine Formation (Late Cretaceous) dinosaurs, 100–117; Two Medicine Formation

nondinosaurian vertebrates, 116–17. *See also* Aptian/Albian Stage; Campanian Stage; Claggett Formation; Cloverly Formation; Eagle Formation; Maastrichtian Stage; Santonian Stage; Turonian Stage; Two Medicine Formation; Virgelle Sandstone

crocodiles, 66; in a cladogram, 14

crossbedding, diagram of, 26

Crow Nation, 45, 66

curation, 148–50

Currie, Phil, 56

Custer, General, 45

Cut Bank, Montana: geology of, 24, 25, 71; in stratigraphic section, 24

cycads, 63, 66

Daspletosaurus, 71, 111–12

dating techniques, 18–19

Deinodon horridus, 44

Deinonychus, 2, 15, 63, 66; *antirrhopus*, 97–98; discovery of, 53; hunting a titanosaur (painting), 67

Deinosuchus, 71

Devonian Period: geological time scale, 17

Diapsids, 13

Dinosaur National Monument, 34, 48

Dinosauria: cladogram, 14, 15

dinosaurs: as term coined by Richard Owen, 53; bipedalism, 45; coprolites of, 42, 43; eggs, embryos, and nests of, 40–43, 56, 142; footprints of, 39–40; gizzard stones of, 43; in cladograms, 14, 15; mummified, 38; preservation of, 1921; skin impressions of, 37–38; study of bone thin sections of, 37; where found in

Montana, 15, 22, 25; why found in Montana, 19. *See also* ornithischian dinosaurs; saurischian dinosaurs

Diplodocus, 15, 57–58, 62, 63, 90; juvenile herd stuck in mudflat (painting), 64–65

Dodson, Montana, 69

Dodson, Peter, 55

dome-headed dinosaurs. *See* Pachycephalosauridae

Douglass, Earl, 47–48

dromaeosaurids within Dinosauria cladogram, 15

Dromaeosaurus albertensis, 123

duck-bill dinosaurs. *See* Hadrosauridae

dunes, 27

dung, dinosaur, 35, 36, 42, 43

Eagle Formation: in stratigraphic section, 23, 69

Edmontonia, 15, 111

Edmontonia rugosidens, 52, 111

Edmontosaurus, 15, 48, 77, 108–9; *annectens*, 132–34

Egg Mountain, 19, 33, 71

eggs, dinosaur, 40–3, 56, 58

Einiosaurus procurvicornis, 2, 56, 71, 108–9

Ekalaka, Montana, 52

Ellis Group: in stratigraphic section, 23

Elmisaurid, 137

embryos, dinosaur, 56

Ethridge, Montana: geology of, 24; in stratigraphic section, 24

Euoplocephalus, 110–11

evolutionary relationships, 13–15

extinction, 63, 79

fake fossils, 11

ferns, 63, 66, 70

field notebooks, 5–8

field numbers, 6–8

field schools, 153

fish: *Acrodus,* 66; *Ceratodus,* 66; *Ceratodus frazieri* (lungfish), 99; *Hybodus,* 66
flood deposits, 26, 32
footprints of dinosaurs, 39–40
Fort Union Group (Paleocene), 25
fossilization process, 19–21, 36
fossils: collection of, 7, 8, 145–46; color of, 37; curation of, 148–50; dating techniques for, 18; donation of, 8, 9, 10; Montana museums that display, 152–53; obtaining permission to collect, 4–5; ownership of, 4, 8, 9; preparation of, 146, 148; preservation of, 8, 19–21, 36, 145–46; value of, 10
Fox Hills Sandstone: in stratigraphic section, 23

Gabriel, Diane, 58
Garbani, Harley, 54, 55, 59
garfish: 31; scales of (illustration), 116
gastroliths, 43
gastropods (snails), 69
geodes, 144
Geological Survey, United States: history of dinosaur expeditions by, 44, 46
geological time scale, 17
Gilmore, Charles, 51–52
ginkgoes, 63, 66
gizzard stones, dinosaur, 43
global positioning system (GPS), use of, 5, 7
glue (for fossil preparation), 145–46
Glyptops (turtle), 66; *pervicax* (turtle), 99; *plicatulus* (turtle), 99
Gobiconodon, 66
Gryposaurus: incurvimanus, 119–20; *latidens,* 15, 56, 101–2
gumbo, 26

Hadrosaur: egg shape of, 41
Hadrosauridae, 101–6, 126, 132–34
Hadrosaurinae, 126
Hadrosaurus, 45; *breviceps* (invalid species), 46; *paucidens* (invalid species), 46
Harlem, Montana, 69
Harlowton, Montana, 47, 69, 71, 74
Harmon, Bob, 59
Harvard University, 55
Harwood, Tom, 49
Hatcher, J. B., 46
Havre, Montana, 25, 71
Hayden, Ferdinand Vandiveer, 44
Hell Creek Formation, 77, 79, 80; aves from, 138; dinosaurs from, 77, 80, 127–38; fossil plants and animals from, 80; in stratigraphic section, 23; nondinosaurian vertebrates from, 138–40; on map of Montana's Mesozoic geology, 22
herds, dinosaur, 33
Hindsdale, Montana: geology of, 25
Hoppers Formation, 77; fossil plants and animals from, 80
Horner, Celeste, 59
Horner, Jack, 56; childhood, 1–2; dinosaurs named by, 56; first dinosaur find, 1; fossil donation, 10
Hybodus (fish), 66
Hypacrosaurus stebingeri, 56, 71, 77, 104–6
"Hypsilophodontid" dinosaurs, 128

invalidated dinosaur species, 11, 45, 46
Isaac, J. C., 45

Jack's Birthday Site, 34, 58
Jenkins, Farish, 55
Joplin, Montana: geology of, 25

Judith River Formation, 71, 74–75; age of, 18; dinosaurs from, 71, 74–75, 117–23; exploration of, 55; exposure of, 25; fossil plants and animals from, 74–75; in stratigraphic section, 23; nondinosaurian fossils from, 124; on map of Montana's Mesozoic geology, 22
Judith River, history of dinosaur hunting near, 44–45
Jurassic Period, 16; climate during, 60; geological time scale, 17; Late Jurassic dinosaurs, 86–92; map of Montana's depositional features during, 62; map of world during, 61. *See also* Morrison formation; ornithischian dinosaurs

Kimmeridgian Stage (Jurassic), 61–63; conditions during, 62; geological time scale, 17; map of Montana's depositional features during, 62; Morrison Formation, 62–63; 86–92; ornithischian dinosaurs of, 87–89; saurischian dinosaurs of, 89–92
Kootenai Formation, 63, 66; in stratigraphic section, 23

lag deposits, 34; on map of Montana's Mesozoic geology, 22
lake deposits, 26, 27
Lambert, Marshall, 52
Latin and Greek terminology, 11–12
leaf impressions, 26
Leidy, Joseph, 44

Leptoceratops, 15, 107–8
Linnaean Classification
  System, 13
Little Big Horn, massacre, 45
Livingston Group:
  dinosaurs from, 125–26;
  on map of Montana's
  Mesozic geology, 22
lizards, 66; in cladogram, 14
Los Angeles County
  Museum, 55
Lothair, Montana: geology
  of, 24, 25

Maastrichtian Stage, 16;
  geological time scale,
  17; Hell Creek
  Formation, 77, 79, 80,
  127-40; Livingston
  Group, 125-26; map of
  Montana's depositional
  features during, 76, 77;
  St. Mary River For-
  mation, 77, 79, 126–27
Maiasaura, 2, 33; coprolite
  of, 42; diet, 43;
  dimensions of nest of,
  43; origin of name, 56,
  82; *peeblesorum,* 102-4
Makela, Bob, 56, 82
Makoshika State Park, 79
Malta, Montana: garfish
  found in *Brachylo-
  phosaurus* skeleton in,
  31, 59; geology of, 25
mammals: cladogram of,
  14; *Gobiconodon,* 66
mapping fossil finds: procedure for, 6, 8, 28, 33
Marias River Formation, 1
Marsh, O. C., 46
measuring a stratigraphic
  section, 7
megalosaur, 66
Mesozoic Era: 16; climate
  during, 60; geological
  time scale, 17; in
  stratigraphic section, 23;
  on map of Montana's
  Mesozoic geology, 22
microsites, 34–35
Microvenator, 63, 66; *celer,* 98
Miners Creek Formation, 77

Mississippian Period:
  geological time scale, 17
Missouri River Breaks,
  history of dinosaur
  hunting in, 44–46
mollusks, freshwater, 63, 66
Montana: climate of
  during Mesozoic Period,
  60; dinosaur-bearing
  formations of, 15, 22–25;
  dinosaurs first discovered
  in, 2, 47, 52, 53, 55,
  56; first dinosaur
  museum in, 52; geology
  of, 19, 22, 23; Highline
  geology, 24–25; history
  of dinosaur expeditions
  in, 44–59; land mana-
  gement agencies, 154;
  map of Mesozoic
  formations in, 22;
  museums in, 151–53;
  repository museums in,
  151–53; state fossil
  (*Maiasaura peeblesaurum*),
  70; stratigraphic section
  of Mesozoic geology, 23;
  Triassic depositional
  features of (map), 61;
  why dinosaurs are found
  in, 19
Montanaceratops, 77;
  cerorhynchus, 127
Morrison Formation (Late
  Jurassic), 62–63;
  description of, 62;
  gizzard stones from, 43;
  in stratigraphic section,
  23; on map of
  Montana's Mesozoic
  geology, 22; ornithi-
  schian dinosaurs from,
  87–89; saurischian
  dinosaurs from, 89–92;
  types of fossils from, 63
mosasaurs, 69
mudstone, 26
Murphy, Nate, 59
Museum of the Rockies, 8,
  9, 57, 151; website, 150
museums, purpose of, 8,
  83, 151

naming dinosaurs, 82–83
Naomicheleys (turtle), 66;
  speciosa (turtle), 99
nests of dinosaurs, 41–43
notebook, field 5–8

Olson, Ken, 58
Ordovician Period:
  geological time scale, 17
ornithischian dinosaurs,
  15; from Cloverly
  Formation (Early
  Cretaceous),93–97; from
  Hell Creek Formation,
  128–35; from Judith
  River Formation (Late
  Cretaceous), 118–22;
  from Morrison Forma-
  tion (Late Jurassic),
  87–89; from Two
  Medicine Formation
  (Late Cretaceous), 101–11
ornithischian dinosaurs
  within Dinosauria
  cladogram, 15
Ornithomimus, 98–99;
  velox, 136–37
Orodromeus, 2, 15; *makelai,*
  56, 101
Osborn, Henry Fairfield, 48
Ostrom, John, 53–54
ownership of fossils, 4

Pachycephalosauridae,
  106–7, 129
Pachycephalosaurus, 77;
  wyomingensis, 129
Pachyrhinosaurus, 71, 77
Palaeoscincus rugosidens, 52
paleontology, definition of, 16
Paleozoic Era: geological
  time scale, 17
pay-to-dig programs, 153
Peabody Museum, 10, 46
Pedilla, Sonja, 58
pelecypods, 69
Pennsylvania, University
  of, 44
Pennsylvanian Period:
  geological time scale, 17
Permian Period: geological
  time scale, 17
permineralization, 37

permits, fossil collection, 4, 154
Phanerozoic Eon, 16; geological time scale, 17
plants, flowering, 63
plaster jackets (for fossils), 146, 147
plesiosaurs, 69
pond deposits: geology of, 26-27
potassium-argon dating, 19
preparation, 146, 148
Princeton University, 8, 10, 47
Prosaurolophus blackfeetensis, 56, 71, 104
Protoceratopsidae, 107-8
pseudofossils, 11, 141-44

Quaternary Period: geological time scale, 17

Rapelje, Montana, 47
recording fossil discoveries, 5-8, 28
registering fossil collections, 9
reptile, sphenodontid: *Toxolophosaurus*, 66
Reptilia, 13; cladogram, 14
Richardoestesia, 137; *gilmorei*, 115
rivers: geology of Mesozoic, 26
Running, He Who Picks Up Stones While, 44
Russell, Charles M., Wildlife Refuge, 79

Saint Mary River Formation, 77, 79; on map of Montana's Mesozoic geology, 22; stratigraphic section, 23, 24
Sampson, Scott, 56
sand dunes: geology of, 27
sandstone, 26
Santonian Stage (Late Cretaceous), 69; Claggett Formation, 69, Eagle Sandstone, 69; Virgelle Sandstone, 69
saurischian dinosaurs: 15; Early Cretaceous, 97-100;

from Cloverly Formation, 97-99; from Hell Creek Formation, 135-38; from Judith River Formation, 122-23; from Morrison Formation, 89-92; from Two Medicine Formation, 111-16; Late Cretaceous, 111-16, 122-23, 135-38; Late Jurassic, 89-92
Sauropelta edwardsi, 63, 96
sauropods within Dinosauria cladogram, 15
Saurornitholestes, 114-15
screen-washing, 146
Scythian Stage (Triassic): geological time scale, 17; map of depositional features in Montana during, 61
Seaway, Intercontinental Cretaceous, 23, 69, 77
Seismosaurus, gizzard stones in, 43
septarian nodules, 143
Shelby, Montana, 1, 24, 25
Silurian Period: geological time scale, 17
Sioux, 45
Sitting Bull, Chief, 45
skeletons: articulated, 30-31; associated, 31-32
skin impressions, 37-38, 47
Smithsonian Institution, 51
snakes, cladogram of, 14
statigraphy, 7, 28
Stebinger, Eugene, 51-2
Stegoceras, 15, 106-7; *validus*, 118-19
Stegosaurus sp., 63, 88-89
Sternberg, Charles H., 45, 46
Strayrer, J. F., 51
stream deposits, geology of, 26-27
Struthiomimus, 115-16
Stygimoloch spinifer, 130
Styracosaurus ovatus, 52, 108
Sue, the *Tyrannosaurus rex*, 10
Sunburst, Montana, 69

swamp deposits, geology of, 26
Synapsida, 13

taphonomy, 28-9
Tenontosaurus, 2, 63, 66; *tilletti*, 93-95
terminology, Latin and Greek, 12
Tertiary Period: geological time scale, 17
Teton Trail Museum, 71
theropod: egg shape, 41; teeth, 116, 123, 138
Thescelosaurus neglectus, 128
titanosaur, 66
Titanosauridae, 99
Torosaurus, 58-59; *latus*, 132
Toxolophosaurus (sphenodontid reptile), 66
Trachydon, 48; *mirabilis*, 44-45
Trexler, Laurie, 55, 82
Triassic Period, 16; geological time scale, 17; lack of dinosaurs from, 15; map of Montana depositional features during, 61; map of world during, 60
Triceratops, 15, 48, 52, 58, 77; first remains discovered, 46; *horridus*, 131-32; *prorsus*, 130-31
trionychids, 117
Troodon formosus, 2, 56, 44, 58, 59, 113-12, 122; egg clutch, 42; nest, 42-43
Tullock Formation (Cenozoic): in stratigraphic section, 23
Turner, Ted, 58
Turonian Stage (Middle Cretaceous): description of, 68-9; map of world during, 68
turtles, 63; in cladogram, 14; *Glyptops plicatulus*, 99; *Glyptops pervicax* , 66; *Naomichelys speciosa*, 66, 99

Two Medicine Formation (Late Cretaceous), 70–71; description of, 70; dinosaurs from, 74, 75, 100–117; expedition history of, 52, 56, 58; exposure of, 24, 71; fossil plants and animals, 74–6; in stratigraphic section, 23, 24; on map of Montana's Mesozoic geology, 22

Tyrannosauridae, 125

tyrannosaurids within Dinosauria cladogram, 15

Tyrannosaurus rex, 2, 15, 16, 57, 58, 77, 78, 135–36; diet of, 43; first specimen discovered, 47–48; Museum of the Rockies specimen, 31

Tyrrell Museum, Royal, 5, 56

Ugly Duckling Site, map of, 6

United States Geological Survey, history of dinosaur expeditions by, 44, 46, 51

University of California, Berkeley, 55

University of Montana, 47

University of Pennsylvania, 55

U.S. 2, 24

U.S. 287, 71

Varricchio, Dave, 58

Virgelle Sandstone: deposition of, 69; exposure of, 24, 25; in stratigraphic section, 23, 24

Wankel, Kathy, 57

Weishampel, Dave, 56

Willow Creek Formation, 77; fossil plants and animals from, 79–80; in stratigraphic section, 23

Wilson, Greg, 59

Winifred, Montana, 74

Yale University, 10, 46, 53

Zephyrosaurus schaffi, 2, 15, 66, 95–96; first skeleton discovered, 55

About the Author

Jack Horner, a native of Shelby, Montana, found his first dinosaur bone when he was eight. He now conducts dinosaur research throughout the world. Since 1982, Dr. Horner has been curator of paleontology at the Museum of the Rockies in Bozeman. He and his research team discovered the first dinosaur eggs in the Western Hemisphere, the first evidence of colonial nesting and parental care among dinosaurs, and the first dinosaur embryos—all in Montana! Author of numerous scientific and popular books and articles, Dr. Horner served as technical advisor for *Jurassic Park, The Lost World,* and *Jurassic Park III.*

Jack Horner
—Museum of the Rockies,
Bruce Selyem photo

We encourage you to patronize your local bookstore. Most stores will gladly order any titles they do not stock. You may also order directly from Mountain Press, either by mail using this order form or by calling our toll-free number, 1-800-234-5308, and charging your order to your credit card. Please have your credit card number ready. Call for a free catalog.

Other Mountain Press titles of interest:

_____Roadside Geology of ALASKA	18.00
_____Roadside Geology of ARIZONA	18.00
_____Roadside Geology of COLORADO	18.00
_____Roadside Geology of IDAHO	20.00
_____Roadside Geology of MAINE	18.00
_____Roadside Geology of MASSACHUSETTS	18.00
_____Roadside Geology of MONTANA	20.00
_____Roadside Geology of NEW MEXICO	16.00
_____Roadside Geology of NORTHERN and CENTRAL CALIFORNIA	20.00
_____Roadside Geology of OREGON	16.00
_____Roadside Geology of SOUTH DAKOTA	20.00
_____Roadside Geology of TEXAS	20.00
_____Roadside Geology of UTAH	18.00
_____Roadside Geology of WASHINGTON	18.00
_____Roadside Geology of WYOMING	18.00
_____Roadside Geology of THE YELLOWSTONE COUNTRY	12.00
_____Roadside History of MONTANA	20.00
_____Roadside History of YELLOWSTONE PARK	10.00
_____Agents of Chaos	14.00
_____Fire Mountains of the West	18.00
_____Geology Underfoot in Central Nevada	16.00
_____Geology Underfoot in Death Valley and Owens Valley	16.00
_____Geology Underfoot in Illinois	15.00
_____Geology Underfoot in Southern California	14.00
_____Glacial Lake Missoula and Its Humongous Floods	15.00
_____Northwest Exposures	24.00

Please include $3.00 per order to cover postage and handling.

Please send the books marked above. I have enclosed $_____

Name_____

Address_____

City/State/Zip_____

☐ Payment enclosed (check or money order in U.S. funds) **OR** Bill my:

☐ VISA ☐ MC ☐ Discover ☐ AE Daytime Phone:_____

Expiration Date:_____Card No._____

Signature_____

MOUNTAIN PRESS PUBLISHING COMPANY

P.O. Box 2399 • Missoula, MT 59806 • Order Toll-Free 1-800-234-5308
E-mail: mtnpress@montana.com • Website: www.mountainpresspublish.com